# Meaning making
# in secondary
# science classrooms

# Meaning making in secondary science classrooms

*Eduardo Mortimer and Phil Scott*

**Open University Press**
Maidenhead · Philadelphia

Open University Press
McGraw-Hill Education
McGraw-Hill House
Shoppenhangers Road
Maidenhead
Berkshire
England
SL6 2QL

email: enquiries@openup.co.uk
world wide web: www.openup.co.uk

and

325 Chestnut Street
Philadelphia, PA 19106, USA

First published 2003

A catalogue record of this book is available from the British Library

ISBN 0 335 21207 7 (pb)   0 335 21208 5 (hb)

Library of Congress Cataloging-in-Publication Data
CIP data applied for

Typeset by RefineCatch Limited, Bungay, Suffolk
Printed in Great Britain by Biddles Limited, www.biddles.co.uk

# Contents

# Foreword

This marvellous book is groundbreaking in several respects. Perhaps the most obvious of these is that it is one of the best accounts we have to date of how sociocultural theory can be applied to classroom practice. Many readers – and perhaps even the authors themselves – might be tempted to leave it at that, but in my view they have actually done much more. Specifically, they go well beyond application – if application is understood as leaving the theoretical tenets being applied unchanged – and make a major contribution to sociocultural theory itself.

As Mortimer and Scott point out, sociocultural theory is concerned with a host of issues, but a major focus is how social discourse gives rise to the development of mental functioning in individuals. This general formulation owes its outlines to figures such as L. S. Vygotsky and M. M. Bakhtin, but it has always been the case that the real work of sociocultural research lies in how to operationalize and expand upon the basic set of claims that constitutes its conceptual core. This is precisely the contribution that the two authors of this volume have made.

My interpretation of the authors' contribution stems from the fact that sociocultural theory makes its most important advances when it is pushed forward by new empirical analyses or concrete applications. Mortimer and Scott have contributed to both, but their insights stemming from concrete applications are particularly noteworthy. We know much more about conceptual categories and ways to collect and analyse data as a result of their insights into such things as the dimensions of instructional discourse. And to add to this impressive achievement, the authors have come up with a set of clear and accessible ideas that will be widely used by people engaged in the day-to-day practice of teaching.

This dual accomplishment makes for a truly exciting volume. Building on the fundamental claim that 'teaching science involves introducing the learner to the social language of school science' (Chapter 2), Mortimer and Scott go on to specify the implications in some very insightful ways. For example, their 'communicative approach' takes us further than existing analyses by outlining two dimensions along which the discourse of teachers and students can be analysed. The four categories generated from the combination of these two

Support for the writing of this article was provided by a grant from the Spencer Foundation. The statements made and the views expressed are solely the responsibility of the author.

dimensions provide genuine insight into types of discourse that have all too seldom been differentiated. For example, the authors' formulation makes it clear that it is essential to recognize discourse that is both 'authoritative' *and* 'interactive', a combination that is often conflated in existing discussions.

Simply by providing clarity on such categorization issues, Mortimer and Scott have made an important contribution. But they do much more. Building on ideas that Mortimer and Scott have had for some time about the 'rhythm' of instructional discourse, they employ these categories in a dynamic analysis that takes us to a whole new level of understanding of instructional discourse. Harnessing the basic categories in their communicative approach, they provide some of the most principled analysis we have to date about how different forms of discourse emerge and then fall away at various points in the learning process and with various types of content. And in Chapter 3 they complement this with an extension of the I–R–E (initiation–response–evaluation) pattern of instructional discourse. In their expanded version, an 'F' (for feedback) also makes an appearance to provide new flexibility in the analytic tool that we can use to understand the discourse of instruction.

In sum, this is a book filled with powerful insights. It provides major new theoretical and methodological insights for scholars concerned with conducting research, and it will be profoundly useful to teachers and others concerned with the actual business of science instruction. It is very hard to pull off both of these tasks in one publication, but Mortimer and Scott have done just that. It is a testament to the power of conceptualization and clarity of writing that they have managed to produce such an admirable work. They have set a new standard for others to follow as we seek to push the field of science education forward in the future.

James V. Wertsch
Department of Education
Washington University, St Louis

# List of figures

# List of tables

# Preface

This book is the result of ongoing collaborative work between Eduardo Mortimer and Phil Scott. We first met in autumn 1992 when Eduardo spent one year studying in the School of Education at the University of Leeds, and over the following ten years have worked together in both Brazil and the UK. During that time we have talked around our many common interests – including football, popular music, cinema and, of course, sociocultural theory – always trying to challenge ourselves to go beyond the main trends in science education and to offer new ways of researching and putting research into practice. This book offers a synthesis of where we have got to at this point in time, and it also provides a point of departure for a new research programme. The major part of the writing of the book was carried out in February and March 2002, when Phil spent a writing leave, working with Eduardo, at the Universidade Federal de Minas Gerais, Belo Horizonte.

# Acknowledgements

The writing of this book was supported by grants from CNPq (Conselho Nacional de Desenvolvimento Científico e Tecnológico – Brazil) and FAPEMIG (Fundação de Amparo à Pesquisa do Estado de Minas Gerais – Brazil).

We acknowledge the institutional support of the Faculdade de Educação da Universidade Federal de Minas Gerais and its Postgraduate Programme. We also acknowledge the support of the School of Education at the University of Leeds, and in particular the generous support of science education colleagues.

Many thanks go to Lynne and Pedro, the two teachers who allowed us into their classrooms, providing the opportunity for us to capture something of their wonderful professional expertise as they worked with their students. Thanks also go to the students who unconcernedly allowed us to listen in to their fascinating discussions.

We would also like to thank our respective colleagues in both Brazil and the UK who have contributed to the ongoing dialogue that has helped to shape our thinking. Particular thanks go to friends and colleagues Carey Jewitt, John Leach, Andy Hind, Maria Lúcia Castanheira (Lalu), Orlando Aguiar Jr and Jim Wertsch for their interest and insightful comments relating to the book.

Finally, we dedicate this book to our families: Regina, Lucas and Ivan, Joan, Julia and Anita, with whom all things start and end.

# 1 Doing and talking school science

There must have been a time when science lessons saw the teacher standing at the front of the room, and the students sitting firmly in their places. A time when the science teacher presented scientific facts to the class, and the students listened. A time when student participation in science lessons was restricted to copying notes from the board. We know that such a time existed because we both experienced it in our own schooling. We also know that such approaches to science teaching still exist in many parts of the world. Nevertheless, things are changing, certainly at the level of science curriculum policy, and in many countries this is matched by changes in science teaching practice.

Today the imperative for science lessons is that they are based on *student-centred* approaches and *active* teaching and learning. Teacher and students are now out of their seats, working alongside each other, involved in a whole range of different kinds of *hands-on* practical activities. Compared with what we experienced at school, today's science lessons are virtually unrecognizable. Things have moved on a great deal in the past 30 years, and rightly so too. We do, however, harbour some concern about these developments.

At present it seems that much of what goes on in science lessons is dominated by thoughts of what activities the students might become involved in. Thus, schemes of work consist of lists of *experiments*, teachers' curriculum guides centre on *activities for the students* and school science textbooks offer colourful double-page spreads with prominent sections offering *things for you to do*. We worry about the under-representation of the scientific story in all of this. To be a little more precise, our concern is that the emphasis on practical activity has served to draw attention away from what we regard as being *the* key feature of *any* science lesson. That is the way in which the teacher orchestrates the *talk* of the lesson, in interacting with students, to develop the scientific story being taught. Practical activities can be interesting, motivating and helpful in getting ideas across, but they cannot speak for themselves. It is only through the teacher's and students' talk 'around the activities' (Leach and Scott 2002) that science teaching and learning can occur. You may consider this to be an obvious, or self-evident, point but we believe that the talk of science lessons has been, and certainly continues to be, somewhat neglected.

## The 'invisible' nature of science classroom talk

It is an interesting paradox that, while few would disagree with our assertion that teacher and student talk is of central importance in any science lesson, relatively little attention is paid to it, either in science teaching circles or in science education research. In relation to science teaching practices, it seems that teacher talk is just something that the teacher gets on with.

What strikes us, as we visit many schools and observe lots of science lessons, are the *differences* in the ways in which teachers interact with their students in talking about the science subject matter at hand. In some science classes, the air is filled with words. The teacher asks questions that prompt student thinking and the students are able to articulate their ideas, presenting different points of view. Sometimes the students are tentative in what they say, at other times more confident, but always willing to speak and to listen. At times the teacher takes a clear lead in talking through the ideas from the front of the room, with the whole class. At other times the students work in small groups and there is a real buzz around the room. On these occasions it is difficult to pick out the teacher as she moves between groups, prompting students and discussing their progress.

We see *other* science lessons where the teacher and students seem to be engaged in a different kind of activity altogether. Here the teacher asks lots of leading questions and the responses from the students tend to be limited to odd words here and there, interspersed in the teacher's delivery. Very often, the teacher proves to be remarkably skilled in this presentational style, but the big problem is that there is very little room for the students in the activity, and many do not speak at all.

Of course, the pictures that we paint here are caricatures, but at the same time we believe that they capture something of the reality of science class-rooms in both the UK and Brazil, and probably worldwide. It would also be true to say that we see *far more* of the second type of lesson than the first. So, why are there these major differences in practice? We are not in a position to answer this question directly, but we believe that a significant contributory factor must be that lots of science teachers adopt the more presentational style, simply because it represents the existing, invisible, taken-for-granted practice of science teaching. One of the aims of this book is to try to do something to address this situation, by bringing the talk of science classrooms to a position of greater prominence, thus rendering it more visible.

## Does science classroom talk matter?

Does it matter whether school students are directly engaged in talking during science lessons? Does it matter whether the teacher's performance is limited to a presentational style of teaching? Our answer to both of these questions is a resounding *Yes!* The reason why it matters is that we see talk as being central to the *meaning making* process and thus central to *learning*.

The whole of this book is focused on exploring this link between talking, meaning making and learning, and draws heavily on various aspects of *sociocultural* theory. At the heart of our approach lies Vygotsky's perspective on development and learning, which maintains that all learning originates in social situations, where ideas are rehearsed between people mainly through talk. As the talk proceeds, each participant is able to make sense of what is being communicated, and the words used in the social exchanges provide the very tools needed for individual thinking.

The individual meaning making step, in this process, sees each participant bringing together the ideas which they already have (their existing points of view), along with those 'new' ideas presented in the talk. It may be that there is no tension between existing and new views, in which case learning progresses easily for the individual. At other times conflicts may arise, and these will need to be resolved if new and existing ideas are to be integrated. Either way, meaning making can be seen to be a fundamentally *dialogic* process, where different ideas are brought together and worked upon. Ultimately the dialogue is always played out in the individual's head, but this can be in the context of solitary musing over ideas, or face-to-face discussion with another person, or through individual reflection on the ideas presented in a book, and so on.

It is with these sorts of ideas in mind that we argue the case for the central importance of teacher and student talk in science classrooms. It is through talk that the scientific view is introduced to the classroom. Talk enables the teacher to support students in making sense of that view. Talk enables the students to engage consciously in the dialogic process of meaning making, providing the tools for them to think through the scientific view for themselves.

## Moving on: a post-constructivist paradigm

Over the past 20 years or so, the *constructivist* movement has had a significant influence upon both the rationale and practice of science teaching. Although the term constructivism is used very broadly, at least two main features seem to be shared by constructivists: first, that learning demands the active intellectual involvement of students; second, that the students' prior knowledge influences subsequent learning of scientific concepts.

The work presented in this book involves a substantial shift in focus away from studies of students' alternative conceptions, and towards the ways in which meanings are developed through language in the science classroom. Is it the case, therefore, that we are engaged in a different project altogether? Not entirely. The sociocultural perspectives, which we present in this book, incorporate the basic tenets of constructivism (certainly the two expressed above), but go much beyond them in developing a fresh view of what is involved in teaching and learning science. We firmly believe that the sociocultural views, which we introduce in Chapter 2, connect very easily with the context of school classrooms, offering a plausible and valuable account of teaching and learning science. Along the way they also allow us to develop new insights and to address questions such as 'why are alternative conceptions resilient to change via instruction?' and 'why are some science topics more demanding than others to learn and to teach?' All of this leads us to consider that what we have to offer here belongs to a *post-constructivist* paradigm, moving on from, but not ignoring, that constructivist programme.

Our work has been influenced by a number of researchers who have focused on the interactions of science lessons, and on the ways in which meanings are developed through talk. For example, Edwards and Mercer (1987), in their book *Common Knowledge*, examine the relationship between the content of lessons and the practical activities and the talk, which constitute them. In *Talking Science: Language, Learning and Values*, Jay Lemke (1990) proposes that learning science involves learning to talk science, and focuses on the question of *how* students learn to talk science through classroom discourse. Ogborn *et al.* (1996), in *Explaining Science in the Classroom*, focus on the ways in which high school science teachers construct and present explanations in the classroom. Most recently, Kress *et al.* (2001), in *Multimodal Teaching and Learning: The Rhetorics of the Science Classroom*, have explored the ways in which teaching and learning in the science classroom go beyond the spoken word to involve a range of different modes of communication.

In Chapter 3, we set out an analytical framework for characterizing the talk of school science, including both teacher–student and student–student interactions. In developing this framework we have drawn on a number of ideas from these books. Our intention, however, is that the framework goes beyond them, offering a more integrated and comprehensive approach to capturing and characterizing the talk of school science. Readers will be able to judge, for themselves, whether or not we have been successful in achieving this aim.

## More talk please!

After we have argued the case that the talk of science lessons has been a neglected area of interest, it may seem rather odd that we can now point to various international initiatives that are currently being taken to *extend* the kinds of talk used in school science.

Thus, in the UK the high-profile document *Beyond 2000* (Millar and Osborne 1999), which sets out an agenda for science education for the new millennium, emphasizes the importance of student debate about science-related social issues. In North America, there is a powerful movement towards 'inquiry-based' science lessons, in which the students work collaboratively on open-ended activities (see, for example, Roychoudhury and Roth 1996). On both sides of the Atlantic, moves are being made to engage students in the patterns of talk, or modes of 'argumentation', that are characteristic of science (see, for example, Driver *et al.* 2000; Kelly *et al.* 2000; Duschl and Osborne 2002).

While we regard all of these initiatives as being valuable, in that they encourage students to become more involved in talk in science lessons, there is a sense in which they target new areas for study, before the challenges of conventional practice have been addressed. In a nutshell, how helpful is it for a science teacher and students to be exposed to the genre of scientific argumentation, when their normal, daily lessons are based on a routine of teacher presentation? Our view is that the priority must be, first of all, to make these existing practices more 'visible', and then to point towards how they might be extended by employing the different kinds of interactions we discuss in Chapter 3.

## The focus of this book

Following the ideas set out in the previous sections, the focus of this book is on the different kinds of interactions between teacher and students (including those between students) in science classrooms and how these can contribute to meaning making and learning. The theoretical frameworks, analytical approaches and practical insights that we present are the products of a programme of research that we have been engaged in for nearly ten years. Our motivation is to come to a better understanding of teaching and learning science in real classrooms, so that we can contribute to their further development.

Central to the book is the analytical framework, which we introduce and illustrate in Chapter 3. We have developed this framework from various aspects of sociocultural theory, which we discuss in Chapter 2, and from our own detailed observations in many science classrooms. The framework provides a comprehensive, and fully integrated, set of theoretical tools for analysing and characterizing the various ways in which the teacher acts to orchestrate the talk of science lessons in order to support student learning.

In Chapter 4, the framework is used to analyse a sequence of science lessons aimed at introducing the conditions needed for the occurrence of rusting. The primary focus of the analysis is on the interventions of the teacher, and interesting 'rhythmic patterns' are identified in the classroom talk as the lessons progress. The analysis here is presented in a very detailed and systematic fashion, in order to illustrate how the framework can be used as a research instrument to provide insights to classroom interactions.

In Chapter 5, a second sequence of lessons, this time focusing on teaching and learning the particulate theory of matter, is analysed. We chose this lesson sequence because it contrasts with the rusting lessons in a number of important ways. First, the subject matter, which involves applying particle theory to explain various physical properties of matter, is much more intellectually demanding. Second, the lessons include a lot of small-group work, which provides the opportunity to focus the analysis more on the student–student interactions. The framework is applied to the lessons in exactly the same way as with the rusting teaching, but the analysis is not reported in the same lesson-by-lesson detail.

In Chapter 6, we draw together the insights gained from the analyses of the previous two chapters, and offer further general discussion of the ways in which the different aspects of the framework fit together. Finally, we present some specific examples of the ways in which we have drawn upon the ideas in this book, both in running professional development programmes with teachers and in relation to planning and evaluating teaching approaches.

We hope that the ideas and approaches introduced in the book will contribute to thinking and practice in three related areas:

1  *Teaching and learning science*: (a) describing and illustrating the diverse range of teaching interactions in science classrooms; (b) demonstrating and exemplifying the ways in which language underpins science learning in the classroom; (c) showing how these ideas can be drawn upon to inform the professional development of science teachers.
2  *Research methodology*: developing a new socioculturally grounded approach to analysing classroom talk.
3  *Sociocultural studies*: expanding the ways in which sociocultural theory can be applied systematically to classroom contexts.

A final point concerns the fact that the classroom episodes and sequences presented in this book are taken from schools in either the north of England or

Minas Gerais state, Brazil, and that the patterns of interaction between teacher and students in the two contexts bear a striking resemblance. This observation sparks the belief that what we have to offer in this book captures something of the fundamental nature of science classrooms, fundamental to the extent that it comfortably crosses the divide between continents.

.

# 2 Teaching science, learning science

In a science master of education class, one of us was discussing with a group of teachers the difficulties students have in learning some science concepts. One of the teachers commented that very often students 'have the concept, but just can't put it into words.' There was much nodding of heads from the other teachers to support this view. However, one teacher challenged the idea, arguing that understanding something means that you can articulate it and that 'we don't have some kind of mysterious "brain waves" running around inside our heads which allow us to think things . . . it's just words, it's just language. If you can't say it, you don't understand it!' This seemed like just the point for us to start talking (and thinking!) about Vygotsky.

At a recent international science education conference, a debate developed about the possible value of sociocultural theory for the field of science education. One of the delegates posed the question, 'What has Vygotsky got to tell us about teaching and learning science?' He then went on to supply his answer, 'Nothing!' Of course, we do not (and did not) agree with this point of view, but nevertheless it would be fair to say that Vygotsky's ideas do *not* offer a ready-made set of tools for analysing the interactions of science classrooms. Indeed, there is no reason why they should. Vygotsky worked on his views of development and learning at the start of the twentieth century, in a context quite removed from that of the contemporary science classroom.

What we believe Vygotsky *does* offer is a very helpful overall framework that directs attention to the key aspects of teaching and learning. The challenge for researchers is to draw on Vygotskian theory as a means of orientation and then to develop, and to fill in, the detail as the theory is applied to a particular context. Thus, for example, Vygotsky draws attention to the primary importance of talk in social situations, as a necessary precursor to individual learning. In this book we examine in detail the teacher and student talk of the science classroom, with a view to characterizing it, and seeing how it might underpin student learning. In carrying out this detailed work, we go beyond Vygotskian theory and draw upon a range of theoretical ideas from the wider sociocultural field. In this way our starting point lies with Vygotsky, but the theoretical underpinnings for our work are much broader in scope.

## What is involved in learning science?

### A Vygotskian perspective: from social to individual

The view of science learning that we start with here, and that informs the whole of this book, is based on a Vygotskian perspective. Lev Semenovich Vygotsky was born in 1896, in Orsha, a town not far from Minsk in Belarus. His father worked as the manager of a local bank and it is clear that Vygotsky's early years were happy and culturally stimulating. His career as a major intellectual figure in the USSR took off in 1924 when, at the age of 28, he made a brilliant presentation at the Second All-Russian Psycho-Neurological Congress in Leningrad, which resulted in him being invited to join the Psychological Institute of Moscow. Vygotsky was a fiercely hard-working and charismatic leader in Russian academic circles. According to Aleksandr Romanovich Luria, his student and the key member of his academic group, 'The entire group gave almost all of its waking hours to our grand plan for the reconstruction of psychology. When Vygotsky went on a trip, the students wrote poems in honour of his journey. When he gave a lecture in Moscow, everyone came to hear him' (Wertsch 1985: 10). Right up to his death from tuberculosis, in 1934, this 'Mozart of psychology' (Toulmin 1978) researched and wrote a body of work that was initially subject to political censorship in the USSR, but that would, some 40 years later, begin to have a huge impact in academic circles in the Western world.

Central to Vygotsky's perspective is the idea that development and learning involve a passage from social contexts to individual understanding[1]* (Vygotsky 1978). What does this mean? Quite simply, we first meet new ideas (new to us, at least) in social situations where those ideas are rehearsed between people, drawing on a range of modes of communication, such as talk, gesture, writing, visual images and action. Vygotsky refers to these interactions as existing on the *social plane*. The social plane may be constituted by a teacher working with a class of students in school; it may involve a parent explaining something to a child; it may involve a group of friends talking in a restaurant. As ideas are rehearsed during the social event, each participant is able to reflect on, and make individual sense of, what is being communicated. The words, gestures and images used in the social exchanges provide the very tools needed

---

* All the superscript numbers included in the text refer to the 'Further notes' appendix, where expanded discussions of key ideas are presented.

for individual thinking. Thus, there is a transition from *social* to *individual* planes, whereby the social tools for communication become internalized and provide the means for individual thinking. This intimate relationship between talking and thinking becomes very apparent when we start 'talking to ourselves' or 'thinking aloud' about difficult, or stressful, problems. It is no coincidence that Vygotsky's most influential book is entitled *Thinking and Speech* (Vygotsky 1934).

From this *sociocultural*[2] perspective, learning is therefore seen as a process of *internalization*. This involves a movement from social to individual, in which the individual plane is formed, retaining a quasi-social nature, since the tools used for thinking (all the different forms of language and other modes of communication) originate in human culture. The Vygotskian view offers a sharp counterpoint to those psychological, or cognitive, perspectives that describe development and learning as being driven mainly by maturational processes from within. According to the sociocultural perspective, if you wish to investigate the ways in which people typically *think* about the world around them, the place to start is to investigate the ways in which they *talk* and communicate about the world. If you are interested, for example, in how learning occurs in science classrooms, then the place to start is to examine the talk and other modes of communication of science classrooms.

**Meaning making: a dialogic process**

The process of learning and development that is being described here is *not* one that involves ideas being *transferred* directly from teacher to student, parent to child or friend to friend. What is involved, for each participant, is an ongoing process of comparing and checking their own understandings with the ideas that are being rehearsed on the social plane. If, for example, there is a significant overlap between the ideas being introduced by a teacher on the social plane and the developing individual understanding of a student, then learning will not be problematic, as the student is able to assimilate what the teacher is communicating to their existing patterns of talking and thinking. Of course, there is always the possibility that the student finds what the teacher is saying unfamiliar, and struggles to relate it to their existing understandings. In such cases, teaching and learning become more demanding.

Either way, the process of internalization always involves working on ideas, it always involves an *individual meaning making* step. Just as a group of cinema-goers may find, on leaving the cinema, that they have arrived at rather different interpretations of the latest David Lynch film (which is probably the intention), so too there is inevitably the potential for a group of students to arrive at different individual understandings of what the teacher has been talking about in class (which is probably not the intention). This process of individual meaning making is fundamental to contemporary views of learning and has been variously referred to in terms of the transformative nature of learning, or learning as reconstruction.

The meaning making process is further complicated by the fact that words, in themselves, do not *carry* unique meanings. On the contrary, words are inevitably polysemous, acquiring different shades of meaning as the context of usage changes. Related to this point, we are reminded of the eminent Swedish psychologist Ragnar Rommetveit, and his story (Josephs 1998) of a married man who gets up early to cut the grass on the lawn of his house. The telephone rings and his wife answers. The caller is one of her friends, who asks, 'Is your lazy husband still in bed?' The wife replies, 'No, he's outside cutting the grass.' A little later, one of the husband's friends calls and asks, 'Is Fred working?' The wife, knowing the friend's intention of going fishing with her husband, as they always do on Saturdays, replies, 'No, he's outside cutting the grass.' The wife's words are exactly the same, but different meanings are *generated* in the different contexts.

Of course, the same principle applies to meaning making in science classrooms, where commonly used scientific terms, such as energy, force, mass, substance, reaction and living, can signify different things for students and teachers. Furthermore, it is certainly not the case that differences in meanings exist only between teachers and students. Anyone who has ever heard a group of physics teachers debating 'what we mean by energy' will appreciate that scientific concepts do not carry unique meanings.

The fundamental point here is that meaning making is a *dialogic* process, which always entails bringing together, and working on, ideas. To develop further this line of argument, we turn now to the work of another Russian scholar, Mikhail Bakhtin, and his circle of students and co-workers.

Mikhail Mikhailovich Bakhtin was a contemporary of Vygotsky. He was born on 16 November 1895, in the town of Orel, about 300 kilometres south of Moscow in Russia. His family was descended from the nobility but no longer owned property at the time of his birth; in fact Bakhtin's father was a bank official who worked in several cities as Mikhail was growing up. Bakhtin finished the gymnasium school in Odessa and entered the historical and philological faculty of the local university in 1913. He soon transferred to Petersburg University, which proved to be a most exciting place at that time, and especially in the areas of Bakhtin's interests. In 1918 Bakhtin finished university and moved to Nevel, a city in western Russia, where he was to meet the members of the first 'Bakhtin Circle'.

As with Vygotsky, the Soviet revolution of 1917 had a huge impact on Bakhtin's life and interests, which shifted from aesthetics and philosophy of religion to the great issues of the day in the Soviet Union. Bakhtin's academic career was not always easy and he often found himself working through times of great hardship. For example, during the Second World War, Bakhtin finished a book devoted to the eighteenth-century German novel. The manuscript was accepted for publication but the only copy of it disappeared during the confusion of the German invasion. 'The only other copy of this manuscript Bakhtin – an inveterate smoker – used as paper to roll his own cigarettes during the dark days of the German invasion' (Holquist 1981: xxiv). Bakhtin died at the age of 80, already an influential thinker both in Russia and in the West. Like Vygotsky, he emphasized the role of social life and of 'the other' in forming individual consciousness. For Bakhtin, existence, language and thinking were essentially a dialogue.[3]

Bakhtin's perspective on the dialogic nature of understanding is nicely summarized by one of his co-workers, who argued that:

> to understand another person's utterance means to orient oneself with respect to it ... For each word of the utterance that we are in process of understanding, we, as it were, lay down a set of our answering words. The greater their number and weight, the deeper and more substantial our understanding will be ... Any true understanding is dialogic in nature.
>
> (Voloshinov 1929: 102)

The key point here concerns the *dialogic* nature of coming to an understanding. On some occasions the dialogic nature of meaning making is obvious, as ideas are, for example, talked through between teacher and student in a social context. Here, an actual dialogue is played out in which there is exploration of ideas between two people. At other times the student may be silent in class, listening to the talk that surrounds them, trying to make sense of what is being said. Here the student is equally engaged in the dialogic process of coming to an understanding, as they struggle to bring together different ideas, silent but none the less engaged in dialogic meaning making. In such ways the process of meaning making is carried out on both social and individual planes.

## Learning the social language of science

Learning science involves being introduced to the concepts, conventions, laws, theories, principles and ways of working of science. It involves coming to appreciate how this knowledge can be applied to social, technological and environmental issues. The basic tools of science, such as the laws and theories, are developed within the scientific community and have been, and continue to be, subject to processes of social validation. This social validation might be operationalized through peer review of academic papers, conference presentations and discussion, debate in scientific journals or the resolution of a controversial issue. A fundamental aspect of all of these processes concerns the extent to which the scientific knowledge under review fits with empirical observations and measurements of the natural world. Science can thus be seen as a product of the scientific community, a distinctive way of talking and thinking about the natural world, which must be consistent with the happenings and phenomena of that world. Learning science therefore involves being introduced to the language of the scientific community.

There are clear and helpful links to be made here with Bakhtin's concept of a *social language*. According to Bakhtin a social language is 'a discourse peculiar to a specific stratum of society (professional, age group etc.) within a given social system at a given time' (Holquist 1981: 430). Thus, for example, the language used by a solid state physicist to talk about the structure of ceramic materials forms part of one social language, while the language used by a potter to talk about the moulding properties of the same ceramic materials is part of another. Using Bakhtin's terms, learning science involves learning the social language of science, or, at least, one form of that social language.

James Wertsch, the eminent American scholar and interpreter of Russian sociocultural psychology, has built upon the Bakhtinian idea of social languages (Wertsch 1991: 93–118) in suggesting that the range of different social languages that individuals gain competence in make up a *toolkit* of ways of talking and knowing. This toolkit can be drawn upon by the individual, as appropriate, in different contexts. Thus, for example, each of the different social languages that are introduced through the school curriculum (relating to science, history, geography or whatever) can be thought of as a tool, offering a distinctive way of talking and thinking about the world. When confronted with a particular question or problem, the individual can thus select the tool best matched to solving that problem.

In addition to the discipline-based social languages listed above, it is important to recognize that, from birth, each one of us is immersed in an *everyday social language*. This is the language that provides the means for day-to-day communication with others, that provides a way of talking and thinking about all that surrounds us. In a strong sense, the everyday social language acts to *shape* our view of the surroundings, drawing attention to particular features and representing those features in particular ways. For example, the way in which we routinely talk about the Sun 'rising and setting' helps to develop a strong view of the Sun moving through space, rather than the Earth spinning on its axis. The informal or spontaneous concepts (Vygotsky 1934) that constitute an everyday social language include many of those views that are referred to as 'alternative conceptions' or even as 'misconceptions' in the science education literature. Notions of 'plants feeding from the soil' and 'energy getting used up' are examples of everyday ways of thinking and talking, which are developed without conscious awareness through immersion in everyday social language.

By making the distinction between *scientific* and *everyday* social languages explicit, it becomes apparent that it is the body of formal concepts of the natural sciences that provides the *alternative* perspective to the omni-present everyday ways of talking and thinking, rather than the other way round (Berger and Luckman 1967). From this point of view, it is hardly sur-prising that the alternative conceptions or misconceptions identified by the science education community are 'robust' and 'difficult to change'. These are not the ephemeral outcomes of the solitary child trying to make sense of the natural world around them, but the tools of an everyday language that continuously acts to define socially, and reinforce, our ways of talking and thinking.

A further important distinction needs to be made between what might be referred to as the social language of *science* and the social language of *school science*. It is clear that there are differences between *real science*, as carried out in various professional settings, and *school science*, as enacted in the classroom. In fact, it is also clear that there are major differences in experimental tech-niques and kinds of knowledge *within* the category of *real science*. School science has its own history of development and is subject to social and political pressures that are quite different from those of professional science. The science taught in schools focuses on some ideas and ways of thinking (as defined, perhaps, by a national curriculum), and not on others. One small example of this is the way in which solids are typically represented in school science as regular arrays of close-packed particles or atoms. How many of the solids around us actually have such a structure? The answer is 'very few', but this is the canonical representation of solids within school science. Given its own peculiar history and content, school science thus constitutes a social language in itself.

## Why can learning school science be *difficult*?

One of the occupational hazards of being a science teacher (and, in our personal experience, particularly of being a physical science teacher) arises when meeting people for the first time, at social functions, and announcing your profession as 'science teacher'. The tell-tale looks of dismay betray what for many people are none-too-positive experiences of learning science, 'Oh, physics was *so* difficult, it never made much sense to me.' Why should this be? Why should learning the social language of school science so often prove to be problematic? We find it useful to think about this question in terms of the differences between *everyday* and *school science* social languages.

Consider, for example, the familiar event of a child dropping a ball and it falling to the ground. Why does the ball fall to the ground? In everyday terms, it might be argued that 'the ball falls, because you let go of it'. Alternatively, the need for any kind of explanation at all might be questioned: 'Why does the ball fall? It just does!' The scientific point of view is rather different. First of all, it is clear that this is an event that merits explanation: 'Why does the ball fall? Now, that's an interesting question!' According to the scientific view, the ball falls because of the action of the gravitational force. So what is involved, for the scientific novice or student, in coming to terms with the scientific explanation for this particular event?

First of all, the student must develop an understanding of the concept of gravity, and central to this concept is the notion of action-at-a-distance, the notion that the gravitational force can act on the ball without being in contact with it. This is an idea that is quite likely to challenge the learner's basic assumptions (sometimes referred to as *ontological assumptions*) about the nature of the physical world (it certainly did so for Newton). If the learner is to accept the concept of gravity, then they must be prepared to believe that the Earth can pull the ball down to its surface, while not being in contact with it. If the science teacher presents these ideas as if they are obvious, then they are doing the learner a disservice. They are far from obvious, and on first meeting they often don't seem to make much sense. The learner is being asked to talk, and think, about the world in a new and rather strange way.

A further fundamental feature of the concept of gravity is that it is a single idea that can be used to explain a wide range of phenomena. Thus, gravity can be used to account for such apparently unconnected physical phenomena as a ball falling to the ground, tides ebbing and flowing around our coast lines and the Moon continuing along its orbit around the Earth. In other words, the concept of gravity is generalizable to a range of contexts. Generalizability is a fundamental characteristic of scientific knowledge, which applies to all scientific theories and concepts, and is sometimes referred to as an *epistemological feature* (relating to the nature of knowledge).

Learning the scientific explanation for the ball falling therefore involves coming to understand the concept of gravity, revising basic assumptions about the nature of the world (that forces can act without contact) and developing an appreciation of the generalizability of the concept of gravity and an understanding of the contexts in which the concept can, and cannot, be applied. Viewed in this way, it becomes clear that there are significant differences between the everyday and school science views for this simple event. These differences give rise to a *learning demand*[4] (Leach and Scott 2002), associated with coming to understand the scientific explanation, that is far from trivial.

Of course, there are many contexts of school science learning where there is considerable overlap between everyday and school science views. For example, basic ideas about the human skeleton are unlikely to differ much between everyday and school science views, although school science will offer extra information regarding structure and function, and a new terminology. It is in these areas of overlap between social languages that teachers regard topics for study as being 'straightforward' and learners think the topic is just 'common sense'.

### Emotion and learning

One of the problems involved in writing a book such as this, which focuses on the day-to-day activities of science classrooms, is that not all aspects of those activities can be addressed. Here, we are interested in meaning making in science classrooms, and particularly in how various patterns of communication can be developed to support student learning. At the same time, we would wish to emphasize that we do *not* underestimate the influence of other factors at work in classrooms and, in particular, we recognize the fundamental importance of the affective and emotional aspects of teacher–student and student–student relationships in the process of teaching and learning science.

Emotions govern the patterns of behaviour of individuals and are connected with ideas and feelings of reward and punishment, pleasure and pain, intimacy and distance, personal advantage and disadvantage. Emotions can, and do, have a part to play in meaning making interactions specifically, and in cognitive orientation more generally. According to Damasio (1994), while biological drives and emotion may give rise to irrationality in some circumstances, they are also essential for many aspects of rational behaviour.

In attending a science class (or any other class for that matter), it has been shown (Arruda *et al.* 2001) that a student might develop any one of four broad types of response to the subject they are learning, and to the teacher who is teaching it. These types of response can be arranged in a simple hierarchy. At one extreme, the student might totally *reject* both the topic and the teacher. This response is likely to be reflected in their behaviour in class, which may involve disrupting, or making fun of, more attentive classmates. Second, the student might engage *passively* with the subject matter of the lesson and sit there quietly, listening to what the teacher has to say. Alternatively, the student might engage *actively* in the lessons, taking part in classroom debates, asking questions and becoming involved in the activities prescribed by the teacher. Still one step further, the student might be actively engaged in class, to the extent of taking the *initiative* in searching for further information and developing ideas beyond those required by the teacher.

In the science classes that we investigate and analyse in this book, it is likely that all four kinds of student engagement will be represented. We are not going to offer any analysis of the emotional responses of the students to the lessons as they are enacted, but we shall return to the students' personal views of some of their experiences during those lessons, at the end of the book.

### Review: what is involved in learning science?

Returning to our original question, we take the view that learning science, in school settings, involves being introduced to the tools and practices of a school science social language, and coming to see how these might be applied to diverse social, technological and environmental contexts. This social language does not exist in isolation from all others. Learning school science takes place against a backdrop of everyday ways of talking and thinking about those same natural phenomena that are addressed in science lessons. It is not so much the case that school science views *replace* everyday ideas as that they provide an *alternative way* of talking and thinking about the natural world. A mature understanding of school science entails the ability to move confidently between everyday and school science views, understanding the similarities and differences between the two and being able to draw on each as the context demands[5] (Mortimer 1995).

> When my daughters were younger, we used to take them to the cinema and they would always ask for cartons of orange juice. Part of the attraction, for them, of the orange juice was to make noises with the straw when they were coming to the bottom of the carton. As the ever-alert parent I would instruct the girls: 'Suck more quietly.' They understood what I meant. Even as a physics teacher I was never tempted to demand: 'Remove air from the straw more quietly, in order that a partial vacuum is created and the external atmosphere drives the orange into your mouths without a sound.'

## What is involved in teaching science?

### Staging the teaching and learning performance

Having introduced and discussed these fundamental ideas about what is involved in learning science, we now turn our attention to considering the implications of all of this for teaching. Our starting point is an obvious one. We would say that teaching science involves introducing the learner to the social language of school science. It is clear that the science teacher is central to this process, as they take on the role of interpreter, or mediator, of the school science social language. Furthermore, following Vygotsky's perspective on learning and development, we would argue that any science teaching must involve three fundamental parts. First, the teacher must make the scientific ideas available on the social plane of the classroom. Second, the teacher needs to assist students in making sense of, and internalizing, those ideas. Finally, the teacher needs to support students in applying the scientific ideas, while gradually handing over to the students responsibility for their use.

Taken together, we see these different parts or phases as constituting a kind of *public performance* on the social plane of the classroom. This performance is directed or *staged* by the teacher who has planned the *script* and takes the lead in moving between the various activities of the lessons. It is a performance that takes place over time (usually over a sequence of lessons) and sees the respective roles of teacher and students changing. At times the teacher is the centre of attention, with the class listening to what they are saying. At other times the teacher encourages students to participate directly in the performance by asking questions about the topic under consideration. At yet other times the teacher seems to play a supporting role in helping the students with particular tasks, possibly as they work in small groups. We now turn to looking at each of these phases in a little more detail, bearing in mind that in Chapters 4 and 5 we shall see how they might appear in the context of two real teaching sequences.

### Introducing and developing the scientific 'story'

The overall aim of the staging process is, of course, to make the scientific point of view, or scientific story, available to the students and to support their learning of it. In referring to the scientific point of view as the 'scientific story' we have followed the lead taken by Jon Ogborn and colleagues, in their influential book *Explaining Science in the Classroom* (Ogborn *et al.* 1996). As we see it, school science offers an account, a kind of story, of familiar natural phenomena expressed in terms of the ideas and conventions of the school science social language.

Thus, for example, in the case of teaching about simple electrical circuits, the teacher might start with ideas relating to the function of the electric current in transporting energy to different parts of a circuit. Central to this story is the scientific concept of charge, which relates to particles of matter that are too small to be seen and that certainly lie outside the bounds of everyday talk and experience. In introducing the scientific story of charges transferring energy, the teacher must therefore first help the students to develop an understanding of what charges are. Perhaps the teacher will start by explaining that the charges are particles, which are parts of the atoms in the components of the circuit. The teacher might then move on to considering what charges can *do*, possibly offering an account based on the ideas that the charges are set in motion when a cell is connected into a circuit, that they transfer energy as they pass through resistances and that they are conserved as energy is transferred. In just such a way, an explanatory 'story' of electric circuits is gradually developed.

Related to this idea of 'building up' the scientific story, Vygotsky (1934) makes the point that scientific concepts do not have a direct relationship with the objects that they refer to in the world: this relationship is always mediated by other concepts. It is also worth pointing out that Vygotsky's notion of scientific concept comes from a term in Russian that could just as readily be translated as *academic concept* or *formal concept*. Unlike the notion of scientific in English, it applies to more than the natural sciences.

Most of the concepts we refer to in science classrooms, such as charge, current and energy, are theoretical entities, which are part of a conceptual system, and meanings are therefore developed for them as they are talked about and used in relation to the other parts of this system. For example, students extend their understanding of the concept of charge as they bring together notions of charge and time to construct the concept of current. Introducing the concept of charge to the talk of the social plane is just the *starting point* for the development of the electric circuit story. Thereafter the teacher has the responsibility for making available, and supporting students in coming to understand, all those ideas about what charges are, what they can do and how these ideas relate to other concepts and to the happenings of the natural world.

In staging the scientific story, the teacher needs to be aware of the existing and developing understandings of the students, being sensitive to the kinds of things the students are saying in class, and drawing upon their knowledge of students' everyday views in the topic area. The challenge for the teacher is then to develop convincing lines of argument to interact and engage dialogically with those existing understandings. Such arguments might involve posing key questions that get to the heart of student uncertainties, the use of particular analogies to support students in developing their thinking, or perhaps setting up conceptual conflict situations to confront different points of view.

A fundamental feature of the way in which the teacher develops the scientific story is that it must be 'persuasive' (Sutton 1996) in character as the teacher seeks to convince the students of the reasonableness of the scientific story, which is being staged on the social plane of the classroom. Any teacher will acknowledge the importance of this persuasive aspect of staging the teaching performance, as students are introduced to the school science social language.

## Supporting student internalization

There is clearly a difference between making the scientific story *available* on the social plane and having students make individual sense of that story. As outlined above, Vygotsky refers to this individual meaning making step as involving the process of *internalization*.

When students are first introduced to a new word or concept in a science class, they may quickly master the teacher's definition of the word, but this is not the end of the learning process, it is just the beginning. Consider, for example, a student being introduced to the word *molecule*, with the teacher explaining that this is one of the 'smallest chemical building blocks of a substance'. The life of this word for the student is only just beginning. By 'life of the word', we mean the history of development of its meanings for the student. As the student first attempts to use the word molecule, it will probably feel like an alien, or foreign, word in the student's mouth. In the next chapter, we examine an episode of teaching and learning, in which a student says, ' . . . that the particles have . . . that the molecules, that the particles have energy and there is space between them'. The student is unsure of the correctness of using the term molecule in this context; she is just beginning to develop meaning for it. From this uncertain start, the student will gain confidence, until she is able to use the word as her own, investing it with her own emphasis of meaning and using it with her own expressive accent (Bakhtin 1934).

This second phase of staging the teaching and learning performance therefore concerns the ways in which the teacher can act to support students as they gradually develop meanings for new scientific concepts, and gain expertise and confidence in using them. Here, Vygotsky (1978) refers to the role of the teacher as being that of supporting student progress in the zone of proximal development (ZPD).[6] The ZPD is a concept developed by Vygotsky that brings together the progress in learning of the individual student with the key role of the teacher in assisting that learning.

Vygotsky developed the concept of ZPD as an alternative approach to measuring the 'ability' of a student in a particular area of learning. Traditionally, such measurements are carried out through a system of formal examining in which each individual student works through a written paper, under strictly controlled examination conditions. The concept of ZPD offers an alternative point of view, where the ability to be measured involves not only what the student can achieve working alone, but also what they might achieve prospectively, with assistance from a teacher or from some other more expert person. This approach to assessing an ability acknowledges the insight that all teachers have about their students: that some have more potential to make progress in particular areas of study than others. This might be expressed in terms such as 'Although they got the same mark in the test, Jane is more capable, and has the ability to go much further than John.' The ZPD thus marks the difference between what a learner can achieve, in a particular topic area, working with and without assistance. So how can the teacher help students to achieve a level of performance that they are not yet capable of achieving alone?

The first point to be made here is that the teacher's interventions to support internalization of the scientific story by students are made *throughout* the teaching sequence. It is not a case of making the scientific story available and *then* helping the students to make sense of it. In this respect we consider that the continuous *monitoring* of students' understandings and *responding* to those understandings, in terms of how they relate to the intended scientific point of view, must be central to the teacher's role. Of course, these processes of monitoring and responding are made more difficult by the fact that the teacher is working not with one student at a time but with a whole class of students, although the interaction is always one-to-one. As the teacher is engaged in these linked processes of monitoring and responding, he or she is probing and working on the 'gap' between an individual student's existing understandings and their potential level of unassisted performance. They are working with the student in the ZPD.

### Handing over responsibility to the student

The final part of the teaching and learning performance involves the teacher providing opportunities for students to try out and practise the scientific ideas for themselves, to make those ideas their own. This step of applying ideas might be first carried out by students, with the support and guidance of their teacher. Initially it is likely that the students will work on familiar situations and problems, before gradually moving on to new and unfamiliar contexts. As the students gain in competence and confidence, the teacher gradually hands over (Wood *et al.* 1976) responsibility to them, recognizing their increased capability for unassisted performance.

This concept of handover follows directly from the Vygotskian conceptualization of learning, as moving from assisted to unassisted performance. As such, the process of handover is a fundamental part of the teaching performance, and needs to be planned and implemented with just as much care and thought as the other two phases. Our experiences in science classrooms suggest that this is not always the case.

## Teaching science: a multimodal process

In science lessons teachers and students do an awful lot of talking. The teacher stands at the front of the room and talks through the scientific point of view, engages students in question and answer sessions, encourages students to work in pairs and to discuss their ideas. Of course, there is more to the average science lesson than just talking: the teacher and students are involved in lots of other kinds of activities. These might include: observing different phenomena through demonstrations and experiments; working with drawings and sketches; using conventional scientific representations, such as diagrams, charts and graphs; imagining, in their mind's eye, things that are either too small or too big to be seen directly; working with physical models of real world artefacts.

Science teachers use whatever communicative resources are available to support the meaning making process, and it is clear that these resources extend beyond spoken and written words. Gunther Kress and colleagues (2001) have made some fascinating analyses of science classrooms, which demonstrate the way in which teachers act to orchestrate a range of communicative resources. In one particularly memorable account, they describe the way in which the science teacher gestures with his hands and arms to demonstrate the circulation of blood, placing a plastic model of a heart over his own body to show where the heart is located and moving between various diagrammatic representations of the circulatory system. In reading such an account, it is absolutely clear that meaning making in science classrooms involves rather more than just talking. The 'multimodal' nature of the interactions on the social plane is inescapable.

Nevertheless, we believe that talk is the *central* mode of communication of the science classroom, as the students are introduced to the social language of school science. Although it would be impossible to find a classroom where *only* verbal language is being used, all the pictures, diagrams, graphics, models, gestures and actions of the teacher that *are* employed achieve their potential for meaning making from the 'flow of language' (Mortimer and Scott 2000) that surrounds them. Our view is that none of these extralinguistic communicative modes can speak for itself, but that it is the teacher and student talk that centrally carries the development of the scientific story, *along with* the action or diagram or model. Furthermore, we believe that talk provides the principal means by which student ideas and understandings can be probed and worked on by the teacher, in supporting meaning making and internalization.

The approach taken in this book to analysing meaning making in science classrooms is therefore one that takes as a principle focus the various patterns and features of teacher and student talk, and looks to examine how this talk interacts with the other modes of communication as the teaching performance unfolds.

## The speech genre of school science

So far, in this section, we have set out our views, based on a sociocultural perspective, of what is involved in teaching science. We wish to conclude by introducing one further idea from Bakhtin, the concept of *speech genre*.[7]

According to Bakhtin, the diversity in kinds and forms of language used by ordinary people is a result not only of the different social languages used, but also of the distinctive patterns of language used in specific contexts and places. Thus, Bakhtin argues that 'each sphere in which language is used develops its own *relatively stable types* of those utterances. These we may call *speech genres*' (Bakhtin 1953). Bakhtin lists the following as examples of speech genres: daily conversation, everyday narration, the brief standard military command, the elaborate and detailed order, the repertoire of business documents, the diverse forms of scientific statements and so on.

The patterns of talk that prevail in school classrooms are very distinctive and, in themselves, constitute a stable speech genre. Where else, other than in the classroom, does one person (the teacher) ask so many questions to which they already know the answer? Where else are words and phrases repeated time and time again during the course of an interaction?

Just as the science student must learn the social language of science, so too the student must come to recognize the speech genre of the school science lesson and learn how to participate in using that genre. In simple terms, if students wish to learn some science then they must also learn how to engage in the various activities of science lessons. This includes, for example, coming to recognize when it is appropriate to offer their point of view and when they are required simply to 'guess what the teacher is thinking'. They must learn when to speak up, and when to remain quiet. As with social languages, we must learn to recognize the contexts in which it is appropriate to use certain speech genres. For example, using the speech genre of school teaching in wider social situations is unlikely to win new friends, and we can certainly remember those occasions when each one of us has needed to be reminded by our respective wives, 'Don't speak to me like a school teacher! I'm not one of your students'.

In this chapter we have outlined our views on the fundamental issues of what is involved in teaching and learning science. These views are based on sociocultural perspectives, and in Chapter 3 we build on the ideas to develop a framework for capturing and characterizing the speech genre of school science.

# 3   Capturing and characterizing the talk of school science

When we started to visit science classrooms to observe, and to videotape, lessons, we met several teachers who firmly believed that they were 'taking into account students' own ideas', because they (the teachers) were 'always asking questions' and 'getting the students to talk'. When we looked at the videotapes, we realized that most of the questions being asked had only one possible answer, and that was the answer required by the teacher. Furthermore, many of the answers came from the same students, those who were sufficiently confident that their answer was correct (or matched what the teacher wanted). When we reviewed and analysed these kinds of interactions with the teachers, during professional development programmes, some of them were very surprised to realize that although their lessons involved lots of questions and answers and much interaction, there was very little probing of, and working with, students' ideas.

Just as there are distinctive ways in which footballers and football pundits talk to each other about football matches, or ways in which Members of Parliament conduct their business in the House of Commons, so too there are distinctive forms and patterns of talk through which teachers and students interact in science classrooms. In this chapter we are interested in demonstrating various ways of identifying, and coming to understand, those forms and patterns of school science talk. Expressed in terms of the Bakhtinian concepts introduced in Chapter 2, we are interested in finding out more about the *speech genre* of school science and the ways in which that pattern of language use supports development of the school science *social language*. In particular we focus our attention on analysing and characterizing the various ways in which the teacher acts to orchestrate the talk of science lessons in order to support student learning.

To address this aim, we have developed a model, or *analytical framework*, that relates to different aspects of teacher–student (and student–student) interactions in science lessons. The framework is based on the sociocultural view of teaching and learning set out in Chapter 2, and has been developed through closely observing, and interpreting, the interactions and activities of science lessons as they are played out in high schools in both England and Brazil. From the outset, we wish to emphasize that the analytical framework is offered both as a tool for thinking about and analysing science teaching after the event, and as a model to refer to, *a priori*, in thinking about the planning and development of science teaching.

## The framework

The analytical framework is based on five linked aspects, which focus on the role of the teacher in making the scientific story available, and in supporting students in making sense of that story. These five aspects are introduced briefly below, and then discussed and exemplified in greater detail in the second part of the chapter. They are grouped, as shown in Figure 3.1, in relation to the teaching *focus*, *approach* and *action*.

### Teaching purposes

The first aspect addresses the *teaching purposes*. It is clear that as a sequence of teaching progresses, different teaching purposes are addressed and thus each purpose relates to a particular phase of a lesson, or sequence of lessons. The teaching purposes that we identify are:

- opening up the problem;
- exploring and working on students' views;
- introducing and developing the scientific story;

ASPECT OF ANALYSIS

| | | |
|---|---|---|
| FOCUS | 1 Teaching purposes | 2 Content |
| APPROACH | 3 Communicative approach | |
| ACTION | 4 Patterns of discourse | 5 Teacher interventions |

**Figure 3.1** The analytical framework: a tool for analysing and planning science teaching interactions.

- guiding students to work with scientific ideas and supporting internalization;
- guiding students to apply, and expand on the use of, the scientific view, and handing over responsibility for its use;
- maintaining the development of the scientific story.

We have developed this list of purposes both from our observations of science lessons in which there are significant and substantive interactions between teacher and students, and from the basic tenets of the Vygotskian perspective on teaching and learning. The key question that is addressed in relation to this aspect of the analysis of classroom teaching interactions is:

> What purpose(s) is served, with regard to the science being taught, by this phase of the lesson?

### Content of the classroom interactions

In science lessons, the multiple interactions between teacher and students relate to a range of content matter, which might include: the scientific story being taught (possibly involving conceptual, technological or environmental issues); procedural aspects of doing science (for example, how to connect up an electric circuit); management and organizational issues (giving instructions about homework; calling for silence while a student offers a comment).

Each of these, and other, aspects of the content of classroom interactions is clearly important for the satisfactory working of science lessons, but here we focus our attention on the content as it relates to the scientific story being taught (the school science subject matter of the lessons). We therefore frame our analysis of the content of the interactions in terms of three categorizations:

- everyday–scientific;
- description–explanation–generalization;
- empirical–theoretical.

The *everyday–scientific* dimension allows identification of the broad social language that is being used by either teacher or students at different points in science lessons. The distinction between *description*, *explanation* and *generalization* is a fundamental feature of the scientific social language. These three categories can be further qualified as being either *empirically* or *theoretically* based (allowing us to distinguish, for example, between an empirically based description and a theoretically based description). The key question addressed in relation to this aspect of the analysis of classroom interactions is:

What is the nature of the knowledge which the teacher and students are talking about during this phase of the lesson?

## Communicative approach

The concept of *communicative approach*[8] lies at the heart of the framework, focusing on the ways in which the teacher works with the students to address the different ideas that emerge during the lesson. We have identified four fundamental classes of communicative approach, which are defined by categorizing the talk between teacher and students along each of two dimensions. The first dimension represents a continuum between *dialogic* and *authoritative* talk, and the second a continuous dimension between *interactive* and *non-interactive* talk. The key question that is addressed in relation to this aspect of the analysis of classroom interactions is:

How does the teacher work with the students to address the diversity of ideas present in the class during this phase of the lesson?

## Patterns of discourse

The fourth aspect, *patterns of discourse,* focuses on the simple, but distinctive, patterns of interaction that emerge between teacher and students during ongoing classroom talk. The key question that is addressed in relation to this aspect of the analysis of classroom interactions is:

What are the patterns of interaction that develop in the discourse as teacher and students take turns in classroom talk?

## Teacher interventions

This final aspect of analysis focuses on the ways in which the teacher intervenes to develop the scientific story and to make it available to all the students in the class. This analysis is based on a scheme (Scott 1997) in which six forms of teacher intervention are identified: shaping ideas; selecting ideas; marking key ideas; sharing ideas; checking student understanding; reviewing. The key question that is addressed in relation to this aspect of the analysis of teaching interactions is:

How does the teacher intervene, at this point in the lesson, to develop the scientific story and to make it available to all of the students?

From this list of five aspects, the second (the content) relates to the content and form of the *school science social language,* while the other four contribute to characterizing the *speech genre* of the science classroom. Although these genre-defining aspects were developed in relation to the teacher's role and actions, they can also be used to characterize the students' activity in the classroom (as we demonstrate later). We now turn to examining and exemplifying each of the five aspects in a little more detail.

## Teaching purposes

In watching any sequence of science lessons aimed at teaching a particular piece of science, the observer might reflect, at any point in the proceedings, on the question: 'What is the teacher trying to achieve here?' In other words, what is the purpose of the teaching during this phase of the lessons, with regard to the science being taught?

In Chapter 2, we set out a view of science teaching as involving the *staging* of a *public performance* on the social plane of the classroom. This staging is directed by the teacher, who has planned the script for the performance and takes the lead in moving between the various activities of the science lesson, and through the various phases of a sequence of lessons (which may cover a number of weeks). Central to this overall performance is the job of introducing and developing the scientific story on the social plane of the classroom. There are, of course, other purposes that are addressed during any teaching sequence, and we summarize these in Table 3.1. As indicated above, we have developed this list of purposes from our own experiences of observing and researching science classrooms and from the theoretical insights provided by sociocultural theory.

The teaching purposes set out here are consistent with Vygotsky's perspective on development and learning, as introduced in Chapter 2, as they chart a passage from teacher and students talking through ideas on the social plane, to the students working with those ideas with teacher support, to the students gradually taking responsibility for their independent working. We also believe that the list of purposes makes sense in terms of making links to what is involved in the day-to-day teaching of science in classroom settings. One point to make clear, however, is that this is a list of teaching *purposes* and *not* an algorithmic recipe to be followed through in planning and implementing a science teaching sequence. Experience tells us that teaching and learning rarely, if ever, follow such preordained patterns.

## Content of the classroom interactions

This aspect of the framework focuses on the substantive content of classroom interactions, and as such needs to be applicable to what both the teacher and

**Table 3.1**   Teaching purposes

| Teaching purpose | Focus |
| --- | --- |
| Opening up the problem | Engaging students intellectually, and emotionally, in the initial development of the scientific story. |
| Exploring and working on students' views | Probing students' views and understandings of specific ideas and phenomena. |
| Introducing and developing the scientific story | Making the scientific meanings (including conceptual, epistemological, technological, social and environmental themes) available on the social plane of the classroom. |
| Guiding students to work with scientific meanings, and supporting internalization | Providing opportunities for students to talk and think with new scientific meanings, individually, in groups or in whole-class situations. At the same time, supporting students in making individual sense of, and internalizing, those meanings. |
| Guiding students to apply, and expand on the use of, the scientific view, and handing over responsibility for its use | Supporting students in applying taught scientific meanings in a range of contexts and handing over (Wood *et al.* 1976) responsibility for using those meanings to the students. |
| Maintaining the development of the scientific story | Providing a commentary on the unfolding scientific story, to help students to follow its development and to see how it fits into the wider science curriculum. |

students say. With this in mind the analysis of content is based on three linked categorizations:

- everyday–scientific;
- description–explanation–generalization;
- empirical–theoretical.

**Everyday–scientific**

The first part of the analysis of content allows us to identify the broad social language used at particular points in science lessons, and is based on Vygotsky's distinction between *everyday* and *scientific* concepts, which was introduced in Chapter 2. To see how this approach might be applied in practice, consider the following exchange between student Robin and his science teacher (Bell 1985):

*Like copying something on a synthesizer*

> Teacher: Plant feeding. It's called photosynthesis. Now what does this word mean . . . literally? What does synthesize mean? Synthesis. If you synthesize something what do you do? Robin?
>
> Robin: Like copying something on a synthesizer.
>
> Teacher: Eh?
>
> Robin: Like copying something.
>
> Teacher: Copy? No!
>
> Student: Making up.
>
> Teacher: To make up. Yes, to make up, to make. If you synthesize something you make something, manufacture something. I suppose your synthesizer is for making music, isn't it?
>
> Robin: No, [in patient tone] it's for copying music.
>
> Teacher: Yes, making music. Tunes.

Teacher and student slip past each other in spectacular fashion as the teacher focuses on the scientific meaning for synthesis and Robin brings to bear his understanding of how music synthesizers work. No doubt, this would be one of those occasions when, after the event, the teacher thinks back to the lesson and wonders what Robin meant by what he said. It might well be one of those occasions when the student concludes that the teacher is talking a different language to himself, and of course the student would be correct in this judgement.

### Description–explanation–generalization

The second part of the analysis takes us into the realm of school science and focuses on three fundamental features of the scientific social language: *description*, *explanation* and *generalization*. Much, of course, has been written about these aspects of the scientific social language, but for the purposes of this framework we define each as follows (see also Mortimer and Scott 2000):

- *Description*: involves statements that provide an account of a system, an object or a phenomenon in terms of its constituents, or the spatiotemporal displacements of those constituents.
- *Explanation*: involves importing some form of theoretical model or mechanism to account for a specific phenomenon.
- *Generalization*: involves making a description or explanation that is independent of any specific context.

A further important distinction is that descriptions, explanations and generalizations can be characterized as *empirical* or *theoretical*.[9] Thus, descriptions and explanations that are based on directly observable properties, or constituents of a system, are characterized as empirical, while those that draw upon entities created through the theoretical discourse of science, as in the case of microscopic particle models, are characterized as theoretical (Mortimer 2000).

### Descriptions

Drawing on the ideas set out above, we now consider in more detail what we mean by empirical and theoretical descriptions. To illustrate these definitions we refer to an episode taken from a Brazilian science lesson in which the students were trying to explain, with the help of their teacher, why a grain of potassium permanganate dissolves spontaneously (without stirring) in water. This phenomenon was clearly visible to the students as the purple colour of the permanganate spread throughout the water (Mortimer and Machado 1996).

An *empirical description* is a statement that provides an account of the phenomenon in terms of observable features. Thus, for example, in observing the purple colour of the permanganate spreading, one student offered the empirical description: 'The colour leaves here and comes over here.'

A *theoretical description* goes beyond the phenomenon by drawing on theoretical entities that are not observable in the phenomenon itself. Thus, in relation to the permanganate dissolving, the teacher posed the following question: 'Well, look here. Are you saying that the particles leave here and come over here?' In the teacher's question, the entities referred to (particles) are not a visible part of the phenomenon and in this sense the teacher's utterance goes beyond the phenomenon and is *theoretical* in nature. At the same time, the question is *descriptive* in that it does not involve a proposed mechanism for explaining the phenomenon.

Generally speaking, the descriptions that are most important in constructing scientific stories belong to the second type (theoretical descriptions), as it is not just the phenomenon that is of interest in science, but the way in which the phenomenon is reconstructed in the light of the available scientific theoretical tools.

### Explanations

The category of explanation refers to statements that establish relationships between physical phenomena and concepts, using some form of model or mechanism to account for a specific phenomenon. The following statements were made by a student, Cristina, in relation to the permanganate episode and provide a typical example of making an explanation:

> Cristina: This here is a liquid, and a liquid doesn't have a definite shape, that's right? So . . . there is movement of particles.

Teacher: There is movement of particles . . .

Cristina: And there is space. So, since there is movement and space between the particles, they [particles] tend not to stay in the same place. They, therefore, mix themselves.[10]

In explaining the phenomenon, Cristina relates the physical state of the liquid to the existence of empty space between particles, and to their movement. These are then given as the conditions required for the particles of permanganate and water to 'mix themselves'. Cristina's statements thus go beyond the description of the phenomenon to establish causal relationships to account for it, first, between the physical state of the liquid and the existence of movement and space, and, second, between these features and the phenomenon of mixing.

As with descriptions, explanations can also be classified as being either empirical or theoretical. Cristina's statements constitute a theoretical explanation, since they draw upon a particle model to explain the phenomenon. An earlier statement made by Cristina in the same lesson provides an example of an empirical explanation, in that it is based on the idea that two bodies cannot occupy the same space, something that can be directly observed:

Cristina: . . . the permanganate would not dissolve, it would occupy its space and the water occupies its own space.

Teacher: . . . its own space . . .

Cristina: Because two bodies cannot occupy the same space . . .

### Generalizations

A generalization goes beyond a description and an explanation in that it is not limited to a particular phenomenon, but expresses a general property of scientific entities, matter or classes of phenomena. One student's statement from the permanganate episode provides a typical example of this. The student asserts: ' . . . the particles have . . . the molecules, the particles have energy and there is space between them.'

Note that this student is not referring to particles, energy and space in relation to a particular phenomenon, but to the properties of particles in general, as part of a particulate model of matter. Generalizations can be descriptive or explanatory in nature, although not for a specific phenomenon such as the 'dissolution of potassium permanganate in water' or for a specific kind of matter such as 'copper sulphate solution', but for classes of phenomena and matter such as 'dissolution' or 'liquid'. In the example just given the generalization is theoretical, as it refers to abstract properties (energy and space) of invisible particles, and the whole statement is based on a theoretical model, the particulate model of matter.

Following the introduction of these categories of description, explanation and generalization, it is worth returning to Cristina's explanation to see how all three are often closely linked in their usage. Cristina's explanation provides a typical example of movement in discourse between empirical and theoretical generalizations. Remember that she started by saying that: 'This here is a liquid, and a liquid doesn't have a definite shape.'

In making this statement, she moved attention from this particular liquid 'here and now' to any liquid, as the use of the indefinite article 'a' implies. What she said, therefore, is an example of an empirical, descriptive generalization, as the shape of liquids is something that can be observed. She used this empirical generalization as a first step to bringing into consideration two other generalizations, no longer empirical but theoretical: 'There is movement of particles . . . and there is space.'

Finally she draws upon these general ideas to make a theoretical explanation of the phenomenon: 'since there is movement and space between the particles, they [particles] tend not to stay in the same place. They, therefore, mix themselves.' The fact that the three categories – description, explanation and generalization – all appear in a single statement from the student provides evidence of how these three categories are frequently closely linked.

## Communicative approach

The concept of communicative approach is central to the framework, in providing a perspective on *how* the teacher works with students to develop ideas in the classroom. This aspect of the framework focuses on questions such as whether or not the teacher interacts with students (taking turns in the discourse), and whether the teacher takes account of students' ideas as the lesson proceeds. In developing this aspect of analysis we have identified four fundamental classes of communicative approach, which are defined by characterizing the talk between teacher and students along each of two dimensions: *dialogic–authoritative* and *interactive–non-interactive*.

### The dialogic–authoritative dimension

When a teacher works with students to develop ideas and understanding in the classroom, their approach can be characterized along this dimension, which extends between two extreme positions: either the teacher hears what the student has to say from the student's point of view, or the teacher hears what the student has to say only from the school science point of view.

We refer to the first position as a *dialogic* communicative approach, where attention is paid to more than one point of view, more than one *voice* is heard and there is an exploration or 'interanimation' (Bakhtin 1934) of ideas. The reader should be aware that the sense in which the word 'dialogic' is used here is rather different from the general principle underlying Bakhtin's perspective, which we set out in note 3 of the appendix.

We refer to the second as an *authoritative* communicative approach, where attention is focused on just one point of view, only one voice is heard and there is no exploration of different ideas. The interaction between teacher and student, from the previous section ('like copying something with a synthesizer'), provides a good example of a teacher taking an authoritative approach in completely discounting the student's interpretation of the word 'synthesis' and focusing entirely on the scientific meaning. Generally speaking, classroom interactions are likely to be less clear-cut than this example, displaying aspects of both dialogic and authoritative functions. The reader should refer to note 3 of the appendix for further information about the origins of the distinction between dialogic and authoritative functions.

### The interactive–non-interactive dimension

An important feature of the distinction between dialogic and authoritative approaches is that a sequence of talk can be dialogic or authoritative in nature, independently of whether it is uttered individually or between people. What makes talk functionally dialogic is the fact that more than one point of view is represented, and ideas are explored and developed, rather than it being produced by a group of people or by a solitary individual.

This point leads us to the second dimension to consider in thinking about the communicative approach: that the talk can be *interactive* in the sense of allowing for the participation of other people, or *non-interactive* in the sense of excluding the participation of other people.

### Four classes of communicative approach

Combining the two dimensions, any sequence of classroom talk can be located on a continuum between *interactive* and *non-interactive* talk on the one hand, and between *dialogic* and *authoritative* talk on the other. We represent this combining of the two dimensions in Figure 3.2.

#### Interactive/authoritative communicative approach

At the start of this chapter we described the surprise of some teachers who worked hard to maintain a great deal of *interaction* with the students in their classes, but came to realize that they often paid little attention to the students' ideas. How is it possible for classroom talk to be *interactive* yet *authoritative*? The following interaction took place between a teacher and students in a Year

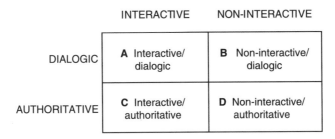

Figure 3.2   Four classes of communicative approach.

8 class in an English high school. The transcript is taken from a series of lessons on energy, and in the previous lesson the students had been working on a practical activity with electric bells.

*Let's just ignore the sparks . . .*

> Teacher:  Do you remember the electric bell?
> Students:  Yes! [in chorus]
> Teacher:  OK! Did any of you notice, did any of you actually hold on to the bell after it had . . . been working? What did you notice?
> Suzanne:  Vibration.
> Teacher:  Well, the arm vibrated, yes. Sound. What else did you notice?
> Tom:  It was loud.
> Teacher:  That's not quite what I'm getting at. Remember the bell. There's the bell [holding up a bell in front of the class]. You did the experiment. If you held on to this bit here where the wires were [indicating], did you notice anything there?
> Jason:  There were sparks there.
> Teacher:  Heat, did you notice some heat?
> Jason:  There were sparks from there.
> Teacher:  There were?
> Jason:  Sparks.
> Teacher:  There were some sparks, yes. Let's just ignore the sparks a minute . . . some heat. There was a little bit of heat there with that one.

In this sequence, it is clear that the teacher is focusing his attention exclusively on the production of heat by the electric bell. He reminds the students about the activities of the previous lesson, and asks, 'What did you notice?' Suzanne replies with a single word, 'Vibration'. The teacher acknowledges her answer, 'Well, the arm vibrated, yes. Sound'. It is clear, however, that this is not what the teacher wants to hear and he moves on, discounting in turn the students' suggestions, 'it was loud' and 'there were sparks'. This is an *authoritative interaction*, where the teacher's sole aim is to arrive at the idea that the bell heats up as it is working.

The teacher's interventions are based on instructional questions for which he has in mind only one answer. If the students do not come up with the required answer, their suggestions are put to one side: 'That's not quite what I'm getting at'; 'Let's just ignore'. The students' contributions are limited to single brief assertions – 'vibration', 'it was loud', 'there were sparks' – made in response to the teacher's questions.

### Interactive/dialogic communicative approach

The interactive/dialogic approach contrasts with authoritative interactions in that here the teacher listens to, and takes account of, the students' points of view, even though these might be quite different from the scientific view. In the context of more interactive teaching, dialogic interactions often occur when the teacher tries to elicit students' views. In the following sequence a teacher is working with a group of 14-year-old students, in a school in the north of England, on the characteristic properties of solids, liquids and gases. The teacher asks the class whether 'solids are hard', an idea that has been suggested by one of the students.

*Other people are desperate to say . . .*

> Teacher: Solids are hard?
> Students: No, no. Soft! [together]
> Teacher: Well, if you say 'no', put your hand up and tell me, give me an
>   example, which would prove an exception to that . . . [the idea that
>   solids are hard].
> Suzanne: Powder's a solid, but you can crush it.
> Teacher: Powder's . . . ?
> Suzanne: . . . a solid but you can still crush it.
> Teacher: Powders aren't particularly hard, yes, if you're talking about hard
>   to the touch. Paul? [who has his hand up]
> Paul: It's . . . cos . . . it's [the powder] got a gas in between, so it's hard.
> Teacher: So you think that all solids are hard?
> Paul: Yeah.
> Teacher: Other people are desperate to say that all solids aren't hard.
>   Martin?
> Martin: Er . . . fabric's soft.
> Students: Yeah . . . yeah . . . [lots of muttering]
> Teacher: Wait. Just a minute. If you're saying things, can you say it to the
>   front, so that we can all share these ideas.

So teacher and class explore the idea that 'solids are hard', with students spontaneously offering points of view in responding to what others have said. The teacher helps to sustain the discussion both by enabling students to contribute ('other people are desperate to say'; 'can you say it to the front?') and by making requests for points of substantive clarification ('give me an example'; 'so you think . . . ?').

Of course, this kind of dialogic interaction can also occur *between students*, especially when working on specific problems in small groups. The following sequence illustrates dialogic interaction as a group of 14-year-old students discusses what happens to the bonding between atoms during cooling, and consequent changes of state (from Wightman 1986).

*How does it happen that bonding comes back?*

> Paul: I mean we're more or less clear how things go from solids to liquids to gases, but not from gases to liquids to solids.
> Jane: The point is, in the gas the bonding has totally gone.
> Paul: So, how does it happen that bonding comes back?
> Jane: I suppose it works vice versa, when it's heated it destroys the bonding, when it's cold . . . you know . . . it remakes it.
> Clare: But how does it remake it? What does it remake it with though?
> Alan: When they're hot they [atoms] vibrate more, so that the bond isn't as strong.
> Paul: Yeah, I know, but they vibrate more and break the bonding and then they finally get to a gas and that's as far as they go . . . but how does it get the bonding back? Ah! . . . when it starts to cool down, they don't vibrate as much.
> Jane: Ah yeah! When they cool down the bonding will be increased so they won't be able to move around as much. That fits in doesn't it?
> Paul: Yeah . . . but the point is, how do we get the bonding back?
> Alan: Slow down the vibrating . . .
> Paul: Slow down the vibrations.
> Alan: I suppose it's ever-present there, but . . . yeah! It hasn't got a chance to like grip, grip them, you know and keep them together. Well, where it slows down, you know, it might get to grips with . . .

The impressive quality of discussion is absolutely striking here as the students move towards the idea that bonding is 'ever-present' and that the strength of bonding increases as substances cool down and atomic vibrations are reduced. A prominent feature of the discourse is the way in which each student makes a contribution in a *tentative* way, inviting further comment and development by others: 'So, how does it happen . . . ?'; 'I suppose it works . . .'; 'but how does it get . . . ?'; 'That fits in doesn't it?'; 'I suppose it's ever-present'. Barnes and

Todd (1995: 161) describe this kind of tentative approach eloquently when they refer to students making a suggestion that 'carries with it the grounds of its own challenge'.

A further interesting feature of this episode concerns the extent to which one of the students (in this case Paul) is able to take on the role of 'teacher'. In the transcript it can be seen that Paul plays a significant part in sustaining the dialogue and keeping it on track, in a way, which is similar to the interventions of the teacher in the previous episode.

### Non-interactive/authoritative communicative approach

Perhaps the best example of a non-interactive/authoritative approach in action is the formal lecture. A friend, Roger, fondly remembers a particular lecture course that he attended as an undergraduate at university. The course was unusual in that he was the only student who opted to take it (it happened to be about quantum mechanical aspects of the theory of materials) and it was taught by a high ranking professor, a Fellow of the Royal Society. The communicative approach was established in the very first lecture. The professor swept into the lecture theatre, which was empty apart from Roger sitting in the middle of the front row, and half-acknowledged Roger's presence before starting to talk expansively, referring to his notes and writing copiously on the board. At the end of the allocated time the professor gathered up his materials and left in silence.

This pattern continued throughout the term, but was interrupted unexpectedly one morning. Midway through the lecture, the quiet atmosphere of study was shattered by the sound of an alarm clock going off. The professor spun round as Roger fumbled in the inside pocket of his jacket for the travelling clock that he used as his time-piece. Eventually, Roger found the alarm-off button and silence was re-established. 'Shall we resume?' asked the professor, and the pair returned to their non-interactive/authoritative way of working. Roger rather enjoyed it that way and it was clear that the professor did too.

### Non-interactive/dialogic communicative approach

At first glance, the notion of a dialogic communicative approach that is also non-interactive appears to be self-contradictory. However, if we remember that a *dialogic* approach is one where attention is paid to more than one point of view (more than one voice is heard), and a *non-interactive* approach excludes the participation of other people, then we can combine these features to imagine a teacher making a statement that addresses the students', or others', points of view but, at the same time, does not call for any turn-taking interaction with the students.

One example of a non-interactive/dialogic teaching intervention occurred at the start of a Year 8 lesson, in which Richard, the teacher, was trying to establish the scientific explanation for why the sides of a plastic bottle collapse inwards when the air inside it is removed. The class is sitting around the front bench listening to the teacher.

*We're going to call that the 'old way' of looking at it*

> Teacher: We need to find out why [the plastic bottle collapses]. We talked about this the other day, when people were saying, 'When it collapses in like that it's because there's something on the inside pulling it.' We're going to call that the 'old way' of looking at it because I want you to think about it by thinking about pressure. I want you to think about it in terms of air pressure. We're going to say that where there's *more* air, there's *more* air pressure. Where there's *less* air, there's *less* air pressure. So the more air there is in a space, the more air pressure there is, and the less air there is in a space, the less air pressure there is.

In this statement Richard first of all reviews the central idea proposed by the students in the previous lesson – 'When it collapses in like that it's because there's something on the inside pulling it' – and then goes on to talk through the scientific explanation in terms of air pressure. In this (non-interactive) way he represents two points of view, taking account of the students' idea (referring to it as the 'old way') and moving on to the scientific point of view (which he later refers to as the 'new way' of explaining).

## Summary: the four classes of communicative approach

The four classes of communicative approach, as introduced in the previous sections, can be summarized as follows:

- Interactive/dialogic: the teacher and students explore ideas, generating new meanings, posing genuine questions and offering, listening to and working on different points of view.
- Non-interactive/dialogic: the teacher considers various points of view, setting out, exploring and working on the different perspectives.
- Interactive/authoritative: the teacher leads students through a sequence of questions and answers with the aim of reaching one specific point of view.
- Non-interactive/authoritative: the teacher presents one specific point of view.

The four classes of communicative approach provide a very useful tool for identifying the different ways in which teachers can work with their students in developing ideas. What this aspect of analysis does *not* tell us is anything about the way in which each communicative approach is actually achieved in the classroom, through the particular patterns of discourse and forms of intervention used by the teacher. For example, if the teacher is engaged in *introducing the scientific point of view* through an *authoritative/interactive* approach, is it the case that they simply 'announce' the science in a careful and clear way, with accompanying questions, so that the science content is unambiguously communicated? The answer to this question is, inevitably, no. Any form of classroom teaching entails a rather special social event in which the teacher is trying to achieve a shared understanding of the subject matter with thirty or so adolescents. These circumstances lead to distinctive patterns and forms of language use by the teacher and these are what we turn our attention to now.

## Patterns of discourse

Let us start by returning to the first episode referred to in the previous section: 'Let's just ignore the sparks'. This episode was presented to illustrate an interactive/authoritative communicative approach, but if we examine the transcript in a different way, it is possible to identify a distinctive pattern of interaction in the talk, as the teacher works his way towards the idea of heat being produced.

### The I–R–E pattern

As it happens, this pattern of interaction is very common in classrooms and is played out in 'patterns of three' with utterances from teacher–student–teacher. It is referred to as a *triadic* 'I–R–E' interaction (Mehan 1979), where:

- I stands for initiation: normally through a question from the teacher.
- R stands for response: from the student.
- E stands for evaluation: by the teacher.

*Let's just ignore the sparks . . .*

> Teacher: OK! Did any of you notice, did any of you actually hold on to the bell after it had . . . been working? What did you notice?
> Suzanne: Vibration.
> Teacher: Well, the arm vibrated, yes. Sound. What else did you notice?
> Tom: It was loud.
> Teacher: That's not quite what I'm getting at. Remember the bell. There's the bell [holding up a bell in front of the class]. You did

the experiment. If you held on to this bit here where the wires were [indicating], did you notice anything there?

Jason:  There were sparks there.

Teacher:  Heat, did you notice some heat?

The teacher initiates the exchange by posing a question, 'What did you notice?' (initiation). Suzanne replies with a single word, 'Vibration' (response). The teacher acknowledges her answer, 'Well, the arm vibrated, yes. Sound' (evaluation).

It is clear, however, that this is not what the teacher wanted to hear and he moves on to his next question, 'What else did you notice?' (initiation). The I–R–E pattern of interaction is very distinctive in all classrooms, and most authoritative interactions (such as the one above) are played out through an I–R–E pattern.

## The I–R–F and I–R–F–R–F– patterns

An alternative form of triadic discourse occurs when, instead of making an evaluation of the student's response, the teacher gives the student *feedback* or *elaborates* on the student's answer, so the student is supported in developing their own point of view. We refer to this pattern of interaction as an I–R–F (F standing for feedback), rather than an I–R–E form. This pattern of discourse can also occur in a chain of interactions, as an I–R–F–R–F– form, where the elaborative feedback (F) from the teacher is followed by a further response from the student (R) and so on. Such a chain of interactions can be seen in the episode 'Other people are desperate to say' from the earlier section.

*Other people are desperate to say . . .*

Teacher:  Well, if you say 'no', put your hand up and tell me, give me an example, which would prove an exception to that . . . [the idea that solids are hard].

Suzanne:  Powder's a solid, but you can crush it.

Teacher:  Powder's . . . ?

Suzanne:  . . . a solid but you can still crush it.

Teacher:  Powders aren't particularly hard, yes, if you're talking about hard to the touch. Paul? [who has his hand up]

Paul:  It's . . . cos . . . it's [the powder] got a gas in between, so it's hard.

Teacher:  So you think that all solids are hard?

Paul:  Yeah.

Teacher:  Other people are desperate to say that all solids aren't hard. Martin?

Martin:  Er . . . fabric's soft.

The teacher starts with 'give me an example' (initiation). Suzanne states, 'Powder's a solid . . . ' (response) and the teacher repeats the comment to sustain the interaction 'Powder's . . . ?' (feedback). Suzanne responds 'a solid but you can still crush it' (response) and the teacher elaborates on her comment, 'Powders aren't particularly hard, yes, if you're talking about hard to the touch' (feedback). Paul then offers an alternative point of view, 'It's . . . cos . . . it's got a gas in between, so it's hard' (response), and the teacher asks for elaboration, 'So you think that all solids are hard?' (feedback).

By establishing this pattern of discourse the teacher is able to explore the students' ideas. In some of his interventions the teacher simply 'bounces back' the student's words, 'Powder's . . . ?', encouraging the student to continue and thereby helping to sustain the interaction. At other points the teacher asks for substantive points of elaboration – 'So you think that all solids are hard?' – in order that the point of view is clarified. In this way the teacher uses an I–R–F–R–F– pattern of discourse to support a dialogic interaction.

## Teacher interventions

This final aspect of the analytical framework focuses on the ways in which the teacher intervenes to develop the scientific story and to make it available to all of the students in the class. Returning to a previous question, is it the case that the teacher 'just tells' the students 'the correct science' in a clear and unambiguous way? The following episode provides evidence to suggest that this is not quite what happens. The episode takes us back to the example 'We're going to call that the "old way" of looking at it', used in an earlier section. Here, Richard the teacher is trying to establish the scientific explanation for why the sides of a plastic bottle collapse inwards when the air inside it is removed. Richard starts by stating the aim of the lesson, but what happens next?

> Teacher: We need to find out why [the plastic bottle collapses]. We talked about this the other day, when people were saying, 'When it collapses in like that it's because there's something on the inside pulling it.' We're going to call that the 'old way' of looking at it because I want you to think about it by thinking about pressure. I want you to think about it in terms of air pressure. We're going to say that where there's *more* air, there's *more* air pressure. Where there's *less* air, there's *less* air pressure. So the more air there is in a space, the more air pressure there is, and the less air there is in a space, the less air pressure there is.

Richard introduces the scientific point of view, emphasizing the importance of the ideas presented by very noticeably modulating his voice and speaking in a slow, clear and deliberate manner. In fact Richard represents *two* points of view. He achieves this by using reported speech to present to the class the 'old way' of explaining ('when it collapses in like that it's because there's something on the inside pulling it'), and then talks through the 'new' way of explaining based on air pressure. So Richard makes a non-interactive/dialogic presentation, thereby explicitly differentiating between everyday and scientific points of view. The non-interactive approach does not last long, however, because Janey, one of the students, has her hand up to speak:

> Janey: Well, when all the air's been sucked out, it's er . . . there's nothing in there, so you'll have . . . air pressure's pushing the sides of the bottle in.
> Teacher: Which air pressure, Janey?
> Janey: From the outside.
> Teacher: Say that again so people can hear.

Janey's initial explanation appears to be consistent with the scientific point of view (where air pressure provides a resultant force on the outside of the bottle to push the walls in), but Richard is aware of students' everyday reasoning in contexts such as these (where the walls are 'sucked' inwards), and so checks the student's meaning: 'Which air pressure, Janey?' Janey provides the correct response and Richard asks her to repeat this, 'so people can hear'. As it happens, Janey probably has the loudest voice of any student in the class! What Richard is doing here is focusing attention on the scientific point of view, and contriving to have Janey repeat it so that it is made available to all the students in the class. Janey duly takes up her role in this unfolding performance:

> Janey: Well . . . when you suck all the air out . . . there's . . . isn't . . . it's really thin compared to the air outside . . . so it pushes it in.
> Teacher: Right, so you're saying that when we suck the air out of the bottle, there's less air in the bottle, so there's less pressure, less air pressure. And why did the sides push in? What did you say again?
> Janey: Cos there's more air pressure outside.
> Teacher: Because there's more air pressure on the outside pushing it . . .

Janey's repeated explanation differs from what she said initially (and from what the teacher wants). Richard responds by paraphrasing what Janey has said, 'Right, so you're saying . . . ', reshaping Janey's statement into the form of the 'new way' of explaining. Richard then breaks off midway through and returns to Janey with a further key question, 'And why did the sides push in?'

Has Richard forgotten what Janey has just said? Clearly not! The purpose of the question is to break down the performance of the scientific explanation into its two main parts. Having considered the conditions inside the bottle, Richard returns to Janey to rehearse what happens on the outside. In this way the teacher shapes the presentation of the scientific explanation. Finally, Richard returns to his careful and deliberate 'presentational voice' and talks through the full scientific explanation:

> Teacher: That's what we're going to call the new way of looking. The new explanation is that there's two lots of air involved here, not one. There's one lot inside the bottle and there's one lot in this room immediately surrounding it. If we take the air out of the bottle that means there's less air inside the bottle than there was before ... there's now more pressure outside and it pushes the sides of the bottle in.

The episode thus ends with this non-interactive/authoritative statement from the teacher. In fact this is the *fourth* time that the scientific way of explaining has been repeated in about the same number of minutes. The importance of the new idea is emphasized through repetition.

This brief episode provides a compelling illustration of the kinds of intervention made by one teacher in introducing his students to a new piece of science. How the teacher achieves this is quite striking. Rather than simply stating the scientific point of view – 'where there's more air, there's more air pressure' – he engages one of the students in rehearsing the science explanation in front of the class. In this 'performance' the teacher:

- acknowledges everyday views and differentiates these from the school science view;
- adopts a special voice to signal the importance of key passages of talk;
- emphasizes key words in the explanation;
- engages a student to help to perform the scientific explanation;
- checks the student's meaning;
- paraphrases the explanation offered by the student;
- breaks down the scientific explanation into its main parts;
- repeats the scientific explanation;
- summarizes the key ideas.

The episode lasted no longer than a few minutes, but the analysis presented here points towards the sophisticated and skilful way in which the teacher guides the talk to make the scientific explanation available to the students. It is clear that the teacher does rather more than 'just tell' students the scientific explanation.

The kinds of teacher interventions that are identified in this episode all fall into a wider classificatory scheme that was developed (Scott 1997, 1998) through close observation, and detailed analysis, of teacher talk in a number of different science classrooms. The scheme also maps on to the existing literature in this area (see, in particular, Edwards and Mercer 1987; Lemke 1990) and the main forms of teacher intervention identified are: shaping ideas; selecting ideas; marking key ideas; sharing ideas; checking student understanding; reviewing (see Table 3.2). The first three relate to how the teacher acts to introduce and develop the scientific story and the remainder refer to other aspects of staging the teaching performance.

**Table 3.2**   Teaching interventions

| Teacher intervention | Focus | Action the teacher might take |
| --- | --- | --- |
| 1  Shaping ideas | Working on ideas, developing the scientific story | Introduce a new term; paraphrase a student's response; differentiate between ideas |
| 2  Selecting ideas | Working on ideas, developing the scientific story | Focus attention on a particular student response; overlook a student response |
| 3  Marking key ideas | Working on ideas, developing the scientific story | Repeat an idea; ask a student to repeat an idea; enact a confirmatory exchange with a student; use a particular intonation of voice |
| 4  Sharing ideas | Making ideas available to *all* the students in a class | Share individual student ideas with the whole class; ask a student to repeat an idea to the class; share group findings; ask students to prepare posters summarizing their views |
| 5  Checking student understanding | Probing specific student meanings | Ask for clarification of student ideas; ask students to write down an explanation; check consensus in the class about certain ideas |
| 6  Reviewing | Returning to and going over ideas | Summarize the findings from a particular experiment; recap on the activities of the previous lesson; review progress with the scientific story so far |

# The five aspects of the framework

These, then, are the five aspects of the framework. In the next two chapters, we turn our attention to using the framework as a tool for analysing two extended teaching sequences. In applying the framework we hope to achieve three things. Our first aim is to provide further exemplification and illustrations of the various aspects of the framework, as they are applied to classroom situations. Second, we aim to show how the different aspects come together to provide an *integrated* analysis of how classroom talk contributes to meaning making in science lessons. Finally, we hope to demonstrate the power of the framework in uncovering emerging patterns and features in the classroom talk, over the time period of a sequence of lessons.

# 4 From everyday to scientific ideas: a teaching and learning performance

> In my class, I used to be like the circus ringmaster . . . I'd have lots of things going on, practical activities, demonstrations, anecdotes . . . and I'd be the centre of attention. To be honest, I was good at it, the kids enjoyed the lessons, but it was exhausting! I came to realise that I was doing all of the work . . . so I left the platform at the front and found myself down the side of the room, more often . . . listening to what the pupils were saying. I felt I'd not only moved my position physically in the classroom, but also in terms of how I thought of my job as a teacher. Working alongside them more often, discussing things. There's still room for the master of ceremonies but you need to vary it and get more in touch with how the class is thinking about things.
>
> (Science teacher)

In Chapter 3, we introduced and exemplified each of the aspects of our analytical framework. Of course, it is one thing to exemplify aspects of the framework in relation to isolated episodes of classroom activity, and quite another to put those aspects to work in trying to characterize the way in which classroom talk unfolds over an extended period of time, encompassing a whole sequence of lessons. This is the focus for the present chapter: to use the framework to examine, in some detail, the development of a sequence of lessons.

The questions that are of interest to us concern the ways in which the different aspects of the teaching and learning performance become prominent and then recede as the lessons move on. For example, is there any change in the communicative approach taken by the teacher? Is there any link between the communicative approach and teaching purposes? Is there any pattern in the ways in which different communicative approaches are realized through different patterns of discourse and teacher interventions? In short, are there any discernible patterns to the way in which the different aspects of the framework fit together over a sequence of lessons?

The lessons we have chosen to examine here are unremarkable insofar as they were taught in a state, inner-city comprehensive school in the UK. What attracted us to the lessons is that they show a teacher working with her students in an interactive way, with the explicit aim of attempting to *start with the students' everyday ideas*, and to move from those ideas to the scientific point of view. In the following sections we use the framework to present a detailed and systematic analysis of the lesson sequence, not only generating insights into the teaching and learning performance, but also demonstrating the use of the framework as a research tool.

## A teaching and learning performance

### The context

Our observation of the science lessons takes us to a high school in a large city in the north of England. The inner-city location of the school is within a rather rundown neighbourhood, which mostly consists of small terraced houses and low-rise apartment blocks. A lot of the older buildings are in a state of disrepair and many are boarded up. Each evening the owners of the local shops put steel shutters over the shop windows to prevent vandalism and theft. The school itself was built about thirty years ago and is now undergoing further building development and refurbishment. It has a good reputation in the locality. The teachers work hard with their largely working-class students and are committed to providing every opportunity for them to progress in and out of school.

Lynne, the science teacher who taught the lessons, was a senior teacher in the school with close to twenty years teaching experience. By any standards Lynne would be regarded as being a very good teacher, her relationship with the students being characterized by a calm, interested and very caring approach. The particular class that we focus upon was composed of 27 students aged 13–14 years, including a wide spread of student ability, with a significant skew towards the lower abilities. We follow three science lessons, each of one-hour duration, which formed part of a unit of work on chemical reactions. The aim of the lessons was to establish what conditions are essential for the rusting of iron to occur, namely that iron, air and water are needed.

### The overall teaching approach

In the UK, the topic of 'rusting' is often taught in lower high school and a common teaching approach involves asserting that iron, water and oxygen are essential for rusting, then carrying out experimental tests to confirm this point. Typically the tests would involve placing iron pins in test tubes, each of which presents a different set of conditions (iron pin with water without oxygen; iron pin without water with oxygen; iron pin with water with oxygen).

Lynne took a rather different approach. Three weeks prior to the first lesson, at the end of a science class, each student was given an iron nail and a small square of abrasive cloth and instructed to polish up the nail such that 'it shines like new'. Lynne then explained to the students: 'I want each one of you to take your nail home and to put it in a place where you think it will go very rusty, as rusty as possible over the next three weeks.' In the lesson prior to the three rusting lessons, the students brought their nails back to school (all of them remembered to bring them in) and each mounted their nail on a sheet of card with information about where the nail had been placed, and why they had placed it there. Lynne and some of the students then made a display on the science classroom wall with the nails placed in sequence, from least to most rusty.

At the start of the first of the three lessons, Lynne reviewed with the students all the different places where they had placed their nails. She then compiled a list of students' ideas about what had caused the rusting to occur in each of those places. Teacher and students worked on this list to identify any factors that were present in *all* the cases of rusting, this with a view to isolating the conditions *essential* for rusting to occur. Finally, the students designed and carried out test tube experiments (using the approach outlined above) to confirm these as the essential conditions.

### The teaching and learning episodes

We now turn to examining, in some detail, each of the episodes of the lessons as the teaching and learning performance unfolds.

#### The 'nails display'

The 'nails display' was striking. It covered the full length of one wall of the teaching laboratory. At one end of the display, the least rusty nail hardly looked any different from its original pristine condition, while at the other end the nails were heavily rusted. As you might imagine, the *nails activity* created a lot of interest. All of the students in the class had contributed to it, and all seemed really pleased with the outcome. In particular, the nails at either end of the scale were the subject of celebration: 'Look at that, it's hardly gone rusty at all!' [peels of laughter]. Of course, the display also caught the eye of other pupils and teachers coming into the room: 'What's this miss, can we try it please?'

One of the students, Jill, had taken her nail home and put it down in the cellar, 'because most other things have gone rusty there'. Ajay had placed his nail in a bowl of water, in the kitchen, 'because I know that rust forms round the wet substance'. Ajay knew this because, 'Sometimes, when I'm on my bike and stuff, I go in puddles and it rusts up.' Claire had also made the link to experiences with her bicycle:

> Well, I've got a bike you see. I haven't been using it lately and it's starting to go all rusty on the handlebars. And Mum and Dad tried to get it off, you know . . . and I thought well if I left that [the nail] out then, and now that I've put it [the nail] outside and I've left it outside, it's gone rusty.

Returning to some of the ideas introduced in Chapter 2, it is clear that the students addressed the nails activity by drawing on their *everyday* knowledge of rusting. They did not talk and think in terms of conditions essential for rusting to occur (although all of them knew that water was needed), but instead tended to refer to *prototypical* examples of rusting. Thus, if things go rusty in the cellar, so too will the nail. The students' initial thinking did not draw upon the social language of science but was firmly rooted in the everyday domain. A number of the students commented that they had 'never really thought about it before'. They knew about the phenomenon of rusting but had never reflected on it with any awareness.

Referring to the framework, it is evident that the construction and presentation of the nails display addressed a number of *teaching purposes*. The very act of taking a nail home was seen, by the students, as being very unusual for a science lesson. There was a genuine buzz of excitement around the room when Lynne first outlined the plan (taking a nail home . . . wow!). Furthermore, 'taking the nail home' prompted each and every student in the class to think through, and to talk to others about, their ideas concerning the hitherto familiar and taken-for-granted phenomenon of rusting. For all concerned, the most surprising feature of the display was that some of the nails had hardly rusted at all. The universal response to this finding was 'well, why should that be?' In this way the display prompted the students to look over the evidence collected by the whole class and to start to generate their own questions about what they could see. The activity was thus very effective in *opening up the problem* for individual students, prompting both intellectual and emotional engagement, while at the same time enabling the teacher to *explore students' views* about rusting.

There is a sense in which the very act of mounting the nails on pieces of card signalled to the students that they were to regard them in a new and different way: no longer an everyday artefact, but now an object for study in science class. Furthermore, the actual display of the nails in the classroom provided the physical means for the students' ideas (as written on the individual cards) to be made visually available, and therefore to be shared by all participants. The process of sorting the students' individual 'nail cards' and laying them out in sequence according to the extent of rusting constituted the first step in working on the information collected by the students, drawing attention to the relationship between degree of rusting and the conditions existing for each nail. In this way the first steps in *introducing and developing the scientific story* were taken.

*Episode 1. What was it about those places that made the nails go rusty?*
At the start of the first lesson, the students were gathered around Lynne's table at the front of the room. Lynne begins by reviewing where various students had left their nails:

1  Teacher: You put them in some really interesting places. The sort of places you put them – Dawn put hers on a slope outside in the garden, and Matthew, Andrew and Louise also put theirs outside in the garden . . . Now – er – Barry put *his* in a cement hole outside in a wall. Clare put hers near the garage. Jill put hers in a cellar. Now all of those went rusty.

Lynne then collects ideas from the students on what it was about the places selected that made their nails go rusty:

2  Teacher: So – what I want to do – put on the board, is perhaps put down your ideas of what it was about the places that made your nail go rusty. What do you think it was – thinking about the places – that made your nail go rusty?
3  Haley: Damp.
4  Teacher: Damp. Now – we'll put things up first of all, then we'll have a think about them in a minute. Right – so, damp [Lynne writes it on the board]. Yes – Cheryl?
5  Cheryl: Moisture.
6  Teacher: Moisture [writes it on board]. Damp, moisture. Anything else? Gavin?
7  Gavin: I put mine in some mud in the garden.
8  Teacher: What was it about that mud that you think made yours go rusty?
9  Gavin: Cos it were all wet and all boggy.
10  Teacher: Wet – so it was wet again. Wet [writes it on board]. Right – wet. Any other ideas, Matthew?
11  Matthew: Air.
12  Teacher: Air – right you think *air* could actually – right [writes it on board]. Air could make it go rusty. Fiona?
13  Fiona: Condensation might.
14  Teacher: Condensation – right [writes it on board]. Dawn?
15  Dawn: Could it be like – climate like – if it's hot or cold?

16 Teacher: Hot or cold. Do some other people think that hot or cold might be something significant, in making something go rusty? Hot or cold – is that an idea – yeah? Hot. Which? Both of them, or just one?

17 Dawn: Both.

18 Teacher: Haley's saying perhaps cold. Cold? [students mutter] Well, is there anybody who put theirs in a hot place and it went rusty? [mutters] Don't forget you're thinking about where you put your nail – what it was – what *things* in that *place* – were making it go rusty. Yes?

19 Student: Cold.

20 Teacher: Right [adds 'cold' to list on board], have we got anything else it could have been? Anyone that hasn't given me an answer yet? No? Andrew then.

21 Andrew: On me bike – if I scrape me bike and leave it out in the rain, it goes rusty.

22 Teacher: So – what are you saying is making it go rusty then? Which of these things, which is causing it to go . . .

23 Andrew: Rain.

Lynne started by inviting the class to offer their ideas on what 'made your nail go rusty'. Her interactions with Haley, Cheryl, Matthew and Fiona follow a triadic pattern, but instead of evaluating (as in I–R–E), Lynne simply accepts the suggestion made by the student and writes it on the chalkboard. The exchanges with Gavin, Dawn and Andrew are different. Here, Lynne asks each student to *clarify*, or to *elaborate upon*, their idea, thus setting up an I–R–F pattern, and with Dawn it can be seen that this develops into an I–R–F–R–F– chain. Lynne initiates the exchange (initiation). Dawn responds: 'Could it be like . . . ?' (response). Lynne feeds back and asks for elaboration: 'Hot or cold. Do some other people think that . . . ' (feedback). Dawn and Haley respond (response). Lynne seeks further elaboration: 'Well, is there anybody who . . . ?' (feedback). A student responds with 'cold' (response) and Lynne adds this to the list on the board.

Eventually, over half of the students offer ideas. It is noticeable that many of them present their ideas as *possibilities* rather than as necessarily *correct answers*. Thus Fiona suggests that 'condensation might'. Dawn tentatively asks, 'Could it be like – climate like – if it's hot or cold?' Even as Dawn makes this suggestion she invites comment through her hesitant approach. Lynne follows this up by asking for other opinions – 'Do some other people think?' – and drawing attention to individual views: 'Haley's saying perhaps cold . . . ' It is clear that the talk here is *interactive* in nature and is located towards the *dialogic* end of the dialogic–authoritative dimension and this, of course, is consistent with the teaching purpose of exploring the students' views.

At the same time, however, it should be recognized that this interactive/ dialogic communicative approach was not entirely open-handed. As the exchanges proceed, Lynne brings her authority to bear in carrying out a preliminary sorting, or filtering, of ideas. In some cases student views are accepted without comment (air, damp), at other times Lynne selects part of a student answer (wet . . . not boggy), which is then listed. In this way Lynne controls what appears on the chalkboard. The teacher's rhetoric for this classroom activity is that 'the aim is to collect *your* ideas' and that the list of ideas on the chalkboard represents '*your* suggestions'. As we can see, this is not quite the case.

At the beginning of the episode the teacher focused attention on the *places* where the students had placed their nails. She then shifted attention to *what it was* about those places that made the nails go rusty, thereby making a further step in introducing and developing the scientific story. In terms of the five aspects of the framework, the key features of this first episode of the lesson sequence are summarized in Table 4.1.

### Episode 2. Have we actually repeated ourselves?
The students are still seated around the teacher's table, and on the board the list of suggested things needed for rusting reads:

Rain, Damp, Moisture, Wet, Salt, Vinegar, Air, Condensation, Cold, Dark.

Lynne invites the students to look more closely at these suggestions:

1   Teacher: Now – what I'd like you to do first of all is to *look* at these
     suggestions – because – is there anything that some of them actually
     have in common – have we actually *repeated* ourselves with any of the

**Table 4.1**   Episode 1. What was it about those places that made the nails go rusty?

| | |
|---|---|
| Teaching purposes | Exploring and working on students' everyday ideas about rusting. Introducing and developing the scientific story: focusing attention on the *things* needed for rusting. |
| Content | Moving from describing places where rusting occurred to describing the *things* in those places that can *cause* rusting. |
| Approach | Interactive/dialogic (but with some authoritative interventions by the teacher). |
| Patterns of interaction | Triadic (without evaluation) and I–R–F–R–F– chains. |
| Forms of intervention | Reviewing; checking student understanding; selecting and sharing student ideas. |

things that we've got on the board at the moment? Kevin, first of all
then – what d'you think we've repeated ourselves with?

2  Kevin: Erm – rain, damp . . . then cold.

3  Teacher: Rain, damp.

When Kevin suggests 'rain, damp . . . then cold', Lynne ignores 'cold' and
selects 'rain, damp'. A number of students call out 'and cold, and condensation' and Lynne selects, from these suggestions, 'condensation'. At this point
*moisture, condensation, rain, damp* and *wet* are all underlined on the board and
Lynne asks what they have in common. She is searching for the term 'water'.

4  Teacher: . . . what have we got in common perhaps with all the things
   we've underlined. What is it Kevin?

5  Kevin: They're all wet.

6  Teacher: Well – they're all wet – so what do we mean by wet then? Is
   there something else about wet?

7  Students: No – wet [other mutters]

8  Teacher: What *is* wet perhaps?

9  Student: [chorus] Water! [laughter]

10 Teacher: Water! So is that the key thing? Ketan, what do you think?
   Is water the key thing here that's linking all of these . . .

11 Ketan: Yes.

12 Teacher: You've said rain, damp, moisture, wet, oh . . . condensation
   and what I'm asking you is 'what do you mean by that?' So what is the
   common link perhaps?

13 Ketan: S'all different forms of water.

14 Teacher: Water. Yeah? Anyone disagree with that? That sounds
   reasonable? OK, so with all of those things we can link up and say
   that water is important.

In contrast to the previous episode, Lynne starts here by asking the
kind of instructional question – 'what have we got in common . . . ?' – to
which she already knows the answer ('water'). This leads (in turns 4–8) to an
I–R–E–I–R–I– pattern in which the students are required to 'guess what teacher
is thinking' and the students find it amusing when someone eventually hits
upon the acceptable answer. When 'water' is suggested, Lynne seizes the word
and initiates a *confirmatory exchange* with Ketan (turns 10–13), which roughly
speaking has the form:

Teacher: This is the case. Student . . . is this the case?
Student: Yes.
Teacher: So, what is the case?
Student: This is the case.

The exchange serves the two functions of marking a key idea and making it available to all of the class.

In this brief episode the teacher has the clear purpose of reformulating 'condensation', 'moisture' and the other terms as 'water'. In a bid to achieve this aim, she: selects from student responses; poses a series of instructional questions; and initiates a confirmatory exchange with a student. A number of different points of view are represented in the sequence, both directly through the contributions of Kevin and Ketan and indirectly through the views represented by the words 'moisture', 'condensation' and 'rain'. At the same time, it is clear that the talk, although interactive in nature, is controlled by the teacher and is located towards the *authoritative* end of the dialogic–authoritative dimension.

The point is eventually established that rain, damp, moisture, wet and condensation are all 'forms of water'. These conditions, which were initially offered by students as descriptions of particular *places* (for example, a damp shed, condensation under the window), are now recast by Lynne as 'water', a 'key thing', which was present in all of those different places. As the teacher uses the term 'water', which is not tied to any particular location, she continues the process of transforming the language used to describe the process of rusting, moving gradually from the 'here and now' of everyday social language to the generalized scientific perspective (Table 4.2).

**Episode 3. What we've done is . . .**
The list on the board now includes seven items and reads:

> water, salt, vinegar, air, cold, dark, dry

Lynne turns to the class and motions for silence. Her voice takes on a formal tone and as she speaks she looks around and raises a finger to indicate that she does *not* want to be interrupted. The students recognize the gesture and remain quiet and attentive.

**Table 4.2**   Episode 2. Have we actually repeated ourselves?

| | |
|---|---|
| Teaching purpose | Developing the scientific story: identifying water as a *key thing* associated with rusting. |
| Content | Moving from describing 'watery contexts' to identifying the common link, 'water'. |
| Approach | Interactive/authoritative. |
| Patterns of interaction | I–R–E–I–R–I–. |
| Forms of intervention | Selecting, marking, sharing ideas. |

1 Teacher: Right – OK – fine. Think what we've done now. What we've actually *done* is try to draw together the reasons why you think your nails have gone rusty. And we've actually tried to tease out what are the *main* factors.

In reviewing 'what we've done now', Lynne makes a subtle retrospective shift. From the students' point of view, they had been engaged in describing the things in the places where their nails rusted. Lynne now refers *not* to describing things in particular situations but to identifying the 'reasons' and 'main factors' that led to rusting. Lynne continues:

2 Teacher: Maybe, even within this list here [water, salt, vinegar, air, cold, dark, dry], it's just perhaps one or two of *those* that are the really *essential* things – the real things that we need for something to rust.

The idea of 'essential things' is thus introduced to the classroom talk. A scientific view of rusting involves not only knowing that iron, air and water are involved, but also that they are the *essential* things. The concept of essential things thus provides a particular epistemological framing for the scientific knowledge about conditions for rusting. Other conditions (such as the presence of salt) might affect the rate of rusting, but water, air and iron are essential for that process. These ideas are presented, by Lynne, in such a way that there are no invitations to discussion; the students recognize the authoritative nature of her approach and remain quiet. For the moment, the time for dialogue has passed (Table 4.3).

### Episode 4. Are there particular columns where you've ticked everything?
There is a change of organization now as Lynne sets the students an activity in which they consider the locations where nails rusted (cellar, garden, bowl of water and so on) and decide which of the seven listed conditions – water,

**Table 4.3** Episode 3. What we've done is . . .

| | |
|---|---|
| Teaching purpose | Maintaining the development of the scientific story; reviewing progress.<br>Developing the scientific point of view: introducing the concept of 'essential things'. |
| Content | Describing the process in terms of 'essential things'. |
| Approach | Non-interactive/authoritative. |
| Patterns of interaction | No interaction. |
| Forms of intervention | Reviewing, shaping ideas. |

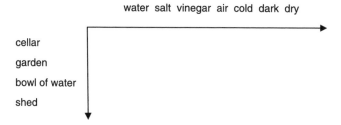

water salt vinegar air cold dark dry

cellar

garden

bowl of water

shed

**Figure 4.1**   The 'things' in each of the places.


salt, vinegar, air, cold, dark, dry – existed in each place. The activity was carried out in small groups and each group was given a large sheet of white paper, on which the students were asked to set up two 'axes', with the seven conditions listed across the top of the paper, and the various locations listed down the side (see Figure 4.1).

The students then marked, with a tick, which of the things existed in each of the locations, the thinking behind the activity being that any things existing in *all* places might be considered 'essential' for rusting. One group finishes the activity and seeks Lynne's attention:

1  Student:  We're finished, miss.
2  Teacher:  You've finished the whole lot? Right, let's have a look. So, now. Looking at everything you've done. Are there particular columns where you've ticked everything? So therefore you think it *must* be that.
3  Student:  Cold?
4  Teacher:  So let me have a look. Well, cold, so it looks like cold could be something.
5  Student:  Air?
6  Teacher:  Air, air.
7  Student:  Vinegar, vinegar's got . . .
8  Teacher:  Vinegar's all crosses so we can discount vinegar, can't we? Right, OK.
9  Student:  Salt's just two.
10  Teacher:  Salt. So perhaps we could discount salt. They're not *essential* factors.
11  Student:  Miss, water.


Lynne starts things off: 'Are there particular columns where . . . ?' The student responds, a little tentatively, 'Cold', and Lynne provides just sufficient feedback to encourage the student to keep going. In this way she works alongside the students, sustaining a dialogic interaction through an I–R–F–R–F– chain. By carrying out the activity the students were able to identify possible essential things. In fact, the activity helps to bring further meaning to the concept of 'essential things' (those things for which there are ticks in all locations). Once again the information was organized and presented in a visual display (as with the 'nails display'), providing a central focus for group discussion.

During the group activity there was much debate about whether certain conditions exist in particular locations (for example, is a shed cold?). A number of groups therefore suggested conditions that were present in *most* but *not all* locations (perhaps because there was still uncertainty about some of these), and these were accepted by Lynne, as the groups reported back at the end of the activity. An ambiguity in the meaning of 'essential things' was thereby created and this was apparent in comments made by students (to the researcher) when asked about the *purpose* of the group activity. Jill, for example, suggested that it was 'to see what mainly made the nails rusty'. Fiona stated that the conditions with most ticks were 'the most common things to make the nails go rusty'. Matthew talked about 'the main things what you need to rust'. The students' talk echoes what was said and done in the activity and does *not* reflect the scientific meaning of 'essential things'. The key aspects of the *group activity* to establish the things essential for rusting are summarized in Table 4.4.

The small groups report back on their findings, and Lynne then summarizes what has been found:

1 Teacher: So, water is everywhere where rusting is taking place and air is also an essential factor. What we have done is to narrow things down to definitely air and water, but it looks as though cold and dark need further testing.

**Table 4.4** Episode 4. Are there particular columns where you've ticked everything?

| | |
|---|---|
| Teaching purpose | Developing the scientific story to identify the *essential things*. |
| Content | Focusing on essential things (empirical description). |
| Approach | Small groups: working around posters.<br>Student–student: interactive/dialogic.<br>Teacher–students: interactive/dialogic. |
| Patterns of interaction | Teacher–students: I–R–F–R–F–. |
| Forms of intervention | Teacher talking with individual groups: checking student understanding; reviewing progress. |

The list on the board reads: air, water, cold, dark. The 'dark' condition is interesting in that only three out of the six groups nominated it. Lynne took the decision that it should be retained as a possible essential factor, perhaps because there had been debate in certain groups about whether dark or light conditions prevailed in particular locations. 'Air' is also interesting, in that it had become an accepted factor without ever having been discussed or questioned by teacher or students. First introduced by Matthew and implicitly accepted by Lynne, it came through the group activity without being challenged (because air is everywhere, even in water).

### *Episode 5. People were talking about the cold*

We have now reached the beginning of the second lesson and Lynne turns the students' attention to the list on the board: air, water, cold, dark. The idea that 'cold is needed for rusting' has become established in the classroom talk, but it is not part of the scientific story. Lynne therefore challenges this notion by referring to personal experiences of holidays in hot places:

1 Teacher: Right, if we want to check . . . that cold, air, water and dark are needed. Now – I thought about this and I was thinking about the first one – the cold, because people were talking about cold. And I suddenly thought about going on holiday somewhere hot – right? Now, how many of you might have been abroad, somewhere *very* hot, like Greece or Spain – places like that? [seven or so students raise their hands]. Right, quite a lot of you. Right, put your hands down. Now, thinking about this I suddenly thought of all the places I'd been to, and I thought, well they're not cold at all – so does that mean that in these places abroad nothing ever goes rusty?

Lynne continues:

2 Teacher: Cos if you think about it – if we're saying it's *cold* that makes things go rusty, then the logic of that is that if you're somewhere *hot* things will never go rusty. Anyone got any comments about that, just put your hand up. Gavin?
3 Gavin: It *can* rust without cold.
4 Teacher: Cold. It can rust without cold – right. Cos d'you get what I'm – *getting* at – Nicola – what does that mean? Gavin is saying that things *can* rust without cold – *should* we then have cold in our list? Is cold essential for rusting then?
5 Students: [chorus] No.
6 Teacher: It's not, is it? No. I even went back and looked through some of my holiday snaps to see if I could *prove* this right. And I got some wonderful photos of some brilliant rusty railings – yes, you can

pass them round. These are pictures from an island in Greece called Santorini – and there's a picture there of some wonderful rusty railings – there's also a picture here of a boat and the *end* you can see has got a lot of rust on it. That seems to me like pretty good proof therefore, that cold – we can cut it out really because it's not absolutely *essential* for rusting.

Lynne thus provides 'proof' for removing cold from the list of possible essential conditions. Lynne's anecdotal account and photographs offer strong visual imagery of railings and boats turning 'wonderfully' rusty in the very hot weather and Lynne draws on these images in presenting a compelling case to suggest that cold is not essential for rusting. From a scientific point of view a number of questions are not addressed. What is meant by 'hot' as opposed to 'cold' conditions? The conditions in Greece are not 'controlled': might rusting have occurred during the cool of the evening, or in the winter? Nevertheless, it appears from the students' responses that they find Lynne's arguments plausible and no one objects when 'cold' is removed from the list. The authoritative way in which Lynne presents her argument adds to its weight, as she makes her initial statement (turn 1), engages Gavin in an I–R–E exchange and then enacts a confirmatory exchange with the rest of the class.

It is worth noting that in developing the argument relating to the cold condition, Lynne refers to the notion of essential things: things can rust without cold, cold is not essential for rusting. In this way the meaning of essential things is rehearsed once more. It has become a feature of the ongoing talk of these lessons, and each time the concept of 'essential things' is referred to, students have the opportunity to redefine and transform their understanding of it.

**Table 4.5** Episode 5. People were talking about the cold

| Teaching purpose | Developing the scientific story: removing 'cold' from the list of possible essential things. |
| --- | --- |
| Content | Focusing on the 'cold' condition (empirical description). |
| Approach | Interactive/authoritative. |
| Patterns of interaction | I–R–E. |
| Forms of intervention | Shaping ideas; marking key ideas. |

### Episode 6. Doesn't matter what you think . . . we're going to set up an investigation

Having removed 'cold' from the list (leaving air, water, dark), Lynne reviews progress and looks ahead to the next step:

1 Teacher: Now that means that we're left with air, water and dark, and what we've got to try to do is to *see* if we can actually prove whether it is *air* on its own, perhaps even *water* on its own, perhaps dark on its own, or a combination of the three which is going to make things go rusty. Now, no matter what you think, doesn't matter what you think, what your ideas are; the point of this afternoon is that we're going to set up an investigation to test that. Right!

In making this statement, Lynne could hardly be more forthright in indicating that a significant point of transition has been reached in the lesson sequence. Whereas at the beginning of the previous lesson, students' views were elicited and dialogue encouraged, Lynne now states that it 'doesn't matter what you think'; scientific experiment, 'an investigation', will provide the basis for testing what makes things go rusty. A further significant pedagogical step has been taken away from everyday views and towards the scientific way of thinking.

Lynne talks through what is involved in designing experiments to test whether the three conditions taken separately, and in combination, lead to rusting. There is some interaction, but the talk is authoritative, with Lynne providing clear instructions of 'what is to be done'. The talk focuses on the scientific logic of how variables can be controlled to test each condition. Groups of students are directed to set up experiments with iron nails in conditions of air alone, water alone, and air and water. Half were left in the dark, with the remainder in illuminated conditions. The controlled conditions for these experiments contrast markedly with the real-life situations of the iron nail activity, a further indicator of progress towards establishing the scientific story.

Table 4.6 Episode 6. Doesn't matter what you think ... we're going to set up an investigation

| | |
|---|---|
| Teaching purpose | Maintaining development of the scientific story: reviewing progress and setting aim for the next step. Developing the scientific story: using scientific experiments to identify the essential things. |
| Content | Epistemological: control of variables as a feature of a scientific experiment. |
| Approach | Non-interactive/authoritative. |
| Patterns of interaction | No interaction. |
| Forms of intervention | Shaping ideas: presenting information. |

*Episode 7. What do the experiments tell us?*
At the start of the third and final lesson, Lynne reviews the activities of the previous two sessions:

1 Teacher: Just to remind you. We were trying to narrow down all the factors we were thinking about that caused rusting – to the absolute – vital ones that were *absolutely* essential. And we'd started to narrow it down by doing the work on the posters – which had left us with four things that we thought were essential – we were left with air, water . . . dark and cold. But we eliminated cold because we realized that if you live in a *hot* country you still have lots of things around you that go rusty. So that left us with these three things – air, water and *dark*. So you set up your experiments last week. Now today – we need to look at these results and see if we *have* narrowed it down any further to the absolute *essential* things that are needed for rusting.

Lynne restates the aim of the lessons, referring to the essential factors as the 'absolute vital ones', and then reviews the steps in the argument from the previous two lessons, making abbreviated references to earlier activities such as 'the work on the posters'. Lynne then invites the class to examine their own experiments. She first directs attention to the test tubes that contained only air, and all agree that no rusting of iron had occurred in those tubes. Lynne next turns to the tubes that contained only water, and these show no rusting, apart from Rebecca's.

1 Teacher: Can I just borrow that tube then, Rebecca, and see if we can think of perhaps *why* – in this particular tube – we *might* have had something go rusty. Think about this carefully. Right – anyone got any ideas – Clare?
2 Clare: Maybe not enough oil, some air might have got in.
3 Teacher: Right – so one point might have been that there – in fact it is quite a *thin* layer of oil – but it still seems to cover it quite well. So it's a good point, but I think, looking at it – what d'you think, Matthew? Do you think there's enough oil on there to stop air getting back?
4 Matthew: No.
5 Teacher: No – well actually Matthew says perhaps there isn't quite enough, so that might have been one point – right? Is there *another* reason though – Rebecca – can you think about your *own* experiment then, and think why?
6 Rebecca: Miss, when I spilt it all out – a lot of it flew out.
7 Teacher: Right – right. So – you put the boiled water in here, and then you dropped the tube and it – no?

8 Rebecca: The oil, Miss.

9 Teacher: You spilt the oil – it dropped out – so that could have been – did any water get out as well?

10 Rebecca: Yeah, it went all over.

11 Teacher: So it was all around. Can anyone think why that might have affected Rebecca's experiment then? Right – Philip do you want to give me an answer?

12 Philip: Y'know when she spilt it? It could have cooled down and let air in.

13 Teacher: Right – I think that's a very good point – and I heard somebody down here – was it Dean? – saying the same thing. Perhaps when it spilt – the air got in.

In this short sequence there is genuine exploration of ideas by teacher and students. As Lynne initially asks the class for ideas about why the iron went rusty, she herself is unaware of what has happened with Rebecca's tube. Lynne engages Claire and then Matthew in I–R–E exchanges to pursue the question and indicates that she does not support the idea of there being not enough oil: 'it's a good point, but I think . . . ' She then turns to Rebecca herself and it becomes apparent that there are very good reasons for the nail rusting! An I–R–F–R–F–R–F chain of interaction develops (turns 5–11) as Lynne asks Rebecca to elaborate on what happened with her experiment. Lynne then looks to the rest of the class and asks 'why that might have affected Rebecca's experiment'. Philip responds, 'It could have cooled down and let air in' and Lynne paraphrases this in repeating it to the class – 'Perhaps when it spilt – the air got in' – leaving out Philip's reference to cooling down. Overall, Lynne employs an interactive/dialogic approach in exploring what happened with Rebecca's experiment, although it is clear that Lynne also uses her authority in moving towards the final acceptable explanation.

**Table 4.7** Episode 7. What do the experiments tell us?

| | |
|---|---|
| Teaching purpose | Exploring students' views: reviewing findings from the students' experiments.<br>Developing the scientific story. |
| Content | Developing an empirical explanation for the unexpected result. |
| Approach | Interactive/dialogic overall. |
| Patterns of interaction | I–R–E and I–R–F–R–F–. |
| Forms of intervention | Reviewing what happened with experiment.<br>Checking student ideas.<br>Selecting ideas: paraphrasing. |

*Episode 8(a). Is that telling us something important?*
Finally, the tubes with air and water present are examined and all of these show rusting (whether in 'dark' or 'light' conditions). Lynne asks what this means:

1 Teacher: So in fact everyone's got their hand up, telling me that with air and water then the nail has gone very rusty. Right – now then. Is that telling us something very important, d'you think? Have we *narrowed* this information down any more? Dawn?

2 Dawn: Well, it means that, means, er, you have to have them both together for the nail to go rusty.

3 Teacher: Right. I think that is an excellent point – and I think it's an excellent way of saying it too. Listen carefully and I'll just re . . . can you just repeat for everyone what you just said?

4 Dawn: Erm, if, if you've got air and water mixed together it's the only time when the nail will go rusty.

5 Teacher: Excellent. You have to have – what you actually said the first time was this – you have to have air and water *together* to make the iron go rusty – and I think that's an excellent way of describing this.

The path from everyday to scientific knowledge, as enacted on the social plane, is now completed, as Lynne engages Dawn in an I–R–E exchange (turns 1–3) and asks her to repeat her statement 'for everyone'. Lynne then paraphrases Dawn's words to arrive at the scientific generalization that 'you have to have air and water *together* to make the iron go rusty', omitting any reference to the nail.

*Episode 8(b). Let's just think back again*
Lynne concludes with a final statement:

Teacher: Let's just think back again. At the start, you were suggesting that it was cold, it was warm, it was dark, it was light, it was acids, or it was – water and air. All those things that were causing rust. That's what we started off thinking. And what we've done *now* – we've now come to the point where you've decided and you've proved in fact that it's *just two* things, with the iron.

Lynne finally refers back to the student thinking at the start of the lessons and draws attention to the difference between the scientific view and the students' initial spontaneous thinking. This is a dialogic statement, in that it represents different points of view. It is also presented in a non-interactive manner, thus providing the first, and last, example in this sequence of lessons of a *non-interactive/dialogic* communicative approach.

**Table 4.8** Episode 8. (a) Is that telling us something important? (b) Let's just think back again

| | |
|---|---|
| Teaching purpose | (a) Developing the scientific story: confirming the essential things. |
| | (b) Maintaining the development of the scientific story: reviewing progress from start. |
| Content | Establishing an empirical, descriptive generalization for rusting. |
| Approach | (a) Interactive/authoritative. |
| | (b) Non-interactive/dialogic. |
| Patterns of interaction | (a) I–R–E. |
| | (b) No interaction. |
| Forms of intervention | (a) Marking key ideas; sharing ideas. |
| | (b) Reviewing. |

### Episode 9. Iron railings next to the sea

In the final activity of the lesson the students, working in small groups, were given a set of cards, each of which described an object in a specific situation. For example, the first card read: 'Ship's bell'. The groups were asked to discuss each card and decide whether or not the object would rust. As the groups worked through the cards Lynne circulated, making occasional comments. Four of the girls worked together on this task.

*Ship's bell*

1 Student: [reading card] An iron bell lost from a ship into the bottom of the deepest part of the Pacific Ocean.
2 Student: Yeah.
3 Student: Yeah, definitely.
4 Student: Cos if you've ever seen, right, them Jaws films, owt like that – you see all metal, all rusty . . .
5 Student: It will rust.
6 Student: It will rust because there's air and how do you think the fishes breathe?
7 Student: Exactly!
8 Student: Yes, so . . .

The students refer initially to images of 'rusty metal' seen on films and then also refer to the presence of air. *Decision*: will rust.

*Lunar rocket*

1 Student: [reading card] Part of a lunar rocket module left on the surface of the moon.
2 Student: No, cos there's no gravity up there.

    3   Student: And there's no air.
    4   Student: No air up there.
    5   Teacher: If there's no air it can't go rusty . . .
    6   Student: Just write rocket.
    7   Teacher: I don't suppose there's any rain there either, is there?
    8   Student: No.

The initial link to gravity is left as attention turns to the lack of air. In turns 5 and 7 the teacher intervenes to feedback, and to elaborate upon, the student ideas. *Decision*: will not rust.

*Handlebars*

    1   Student: [reading card] The handlebar of an old bike with the chrome flaked off.
    2   Students: Yeah, it would, yeah, it would.
    3   Student: Mine's gone rusty.

The students make no explicit reference to essential things; one student refers to her own bike. *Decision*: will rust.

Each of the groups worked through their pack of cards and virtually all of the predictions made by the students were consistent with the scientific point of view.

**Table 4.9**   Episode 9. Iron railings next to the sea

| | |
|---|---|
| Teaching purpose | Guiding students to apply, and expand on the use of, the scientific view, and handing over responsibility for its use. |
| Content | Relating generalized conditions (empirical description) for rusting to specific situations. |
| Approach | Group work: interactive/dialogic. |
| Patterns of interaction | As teacher interacts with groups: I–R–F–R–F–. |
| Forms of intervention | Teacher talking with individual groups: checking student understanding; reviewing progress. |

## The rhythm to the classroom discourse

You know, I watched the lesson. I watched what she was doing and I couldn't figure it out. She moved easily from one activity to the next and the kids were involved, making points, asking questions and following everything she was saying. It looked so effortless and easy, but I know she's a real expert. I just couldn't put my finger on what it was she was doing that made it all fit together so well.

(Student at the start of a science initial teacher training course)

In the previous sections, we examined in some detail the way in which one teacher *talked* her way through a teaching performance, interacting with the students in different ways. We believe that this kind of analysis is very important in focusing attention on the ways in which the teacher addresses and achieves certain aims and purposes, by drawing on and developing different communicative approaches, interventions and patterns of discourse. However, one danger of this kind of micro-analysis is that of becoming lost in the detail of individual episodes and failing to see the broader picture of the way in which the teaching develops over the whole string of episodes that constitute a teaching sequence. We therefore return now to the summary boxes for each episode with a view to 'figuring out' just what it was that this particular teacher was doing over the course of the three lessons on rusting. First, we examine the *content* of the classroom talk.

### The changing *content* of the classroom talk

What can we say, looking back over the previous account and analysis, about the way in which the *content* of the classroom talk changed during the lessons? The lessons were planned to start with the students' own ideas and so, initially, the students were encouraged to describe the *places* where their nails turned rusty. By the end of the lessons, rusting was being talked about as a process involving the *essential things* iron, water and air. In between, the content of the classroom talk, as guided by the teacher, went through a *progressive transformation*, which might be summarized as follows:

- *from* talk about places where rusting occurred;
- *to* possible things in those places that might cause rusting;
- *to* identifying things essential for rusting;
- *to* confirming things essential for rusting using scientific experiments.

Looking at this sequence, there is a sense in which the staging of the scientific story involved a process of gradual *decontextualization* as the teacher guided the classroom talk away from descriptions of rusting events in specific places and towards the statement of a scientific rule applicable to *any* place or situation. Of course, one of the key features of scientific knowledge is that it is generalizable, that it can be applied to a range of different contexts. It is therefore likely that any teaching approach that starts with students' ideas will involve some similar kind of transformation in content moving from the here-and-now of everyday views to the generally applicable statements of science. This kind of process might be better referred to as one of *recontextualization* (in the sense of moving from everyday to scientific worlds), rather than decontextualization (which seems to suggest ending up with knowledge that cannot be linked to *any* context).

A significant point to bear in mind here is that throughout this shift, from everyday to scientific views, the content remains focused on an *empirical–descriptive* account. Thus, at all points in the lessons the talk involved *describing* what is needed for rusting in terms of *directly observable things* (iron, water, air). The shift from everyday to scientific did *not* involve, for example, a move from describing to explaining. An explanation of the rusting process was not part of this particular teaching sequence. By way of contrast, however, in the next chapter we investigate a teaching sequence where the scientific story *does* involve developing explanations that draw on theoretical concepts.

What can we say about *how* this progressive transformation in content was achieved by the teacher? Was it simply a case of the teacher *presenting* (in a non-interactive/authoritative approach) a series of new 'scientific facts' for the students to consider? Was there any shape or pattern to the way in which the teacher guided the development of the scientific story? We start to address these questions by focusing on the changes in communicative approach observed during the three lessons.

## A pattern of changes in the *communicative approach*

When we look back over the analysis of episodes in the previous sections, one striking feature that emerges is the *pattern* in communicative approaches used by the teacher. Thus, in the first three episodes of the sequence, it can be seen that the communicative approach moves through a 'cycle' of:

> (Interactive/dialogic)  –  (Interactive/authoritative)  –  (Non-interactive/authoritative)

This same cycle of approaches is then repeated through the rest of the episodes of the teaching sequence.

In the *first* cycle (episodes 1–3) the teacher: prompted discussion about what might cause rusting in the various nail locations (I/D); interacted authoritatively to identify water as a common factor (I/A); presented a summary of progress (NI/A).

In the *second* cycle (episodes 4–6) the teacher: set up a group activity for students to identify and discuss possible things essential for rusting (I/D); interacted authoritatively to remove the 'cold' condition (I/A); reviewed progress and outlined the next 'experimental' step (NI/A).

In the *third* cycle (episodes 7–9) the teacher: discussed the experimental findings with the students (I/D); interacted authoritatively to confirm the essential things (I/A); summarized overall progress and outcome, referring to both the initial everyday views and the scientific view (NI/D).

In each cycle a similar pattern of activity is apparent:

- teacher and students (or student pairs or groups) *explore* ideas of concern at that point in the development of the scientific story (I/D);
- teacher intervenes to *work on* (through shaping, selecting and marking ideas) some aspect of the content, with a view to developing further the scientific story (I/A);
- teacher *reviews* progress in developing the scientific story, summarizing key points and looking ahead to the next steps (NI/A).

### A fundamental 'rhythm' to developing the scientific story

In this way the classroom talk falls into a kind of rhythm around the repeat step of *explore, work on, review*. In our view the teaching rhythm identified here has much to recommend it for supporting learning in a science classroom context. Why do we believe this to be the case? In the following sections we examine each of the three steps in more detail.

#### Interactive/dialogic approach: 'explore'

The question of why interactive/dialogic talk is important in developing a science teaching sequence is not one that can be satisfactorily addressed in general terms. It depends on what the teacher is trying to do, or the teaching purposes, at specific points in the sequence.

Thus, in the very first episode of the rusting lessons, the teacher invited the students to put forward their views on what it was about the places where they had placed their nails that caused them to rust. Through this activity the teacher was able to learn a great deal about the students' spontaneous views on rusting and to identify a number of issues to be addressed in subsequent teaching. Here we have an example of an interactive/dialogic approach being used to *explore students' ideas*.

In the second cycle of the sequence, the students were involved in a small-group activity to identify and discuss things that might be essential for rusting. Here the teaching purpose was different. The students were given the opportunity to work on their developing understandings, to talk and think through their ideas. The teacher monitored this process, intervening as appropriate to provide help in supporting individual meaning making and internalization. This takes us back to Chapter 2 and to the fundamental principle that developing understanding is a *dialogic process*. If the aim of teaching is for students to develop an understanding of some topic, then those students must engage in some form of dialogic activity, and one way in which this might be achieved is through discussing ideas in a small group. However realized, each student needs the opportunity to engage with the new ideas being taught. Returning to the words of Voloshinov, from Chapter 2:

> To understand another person's utterance means to orient oneself with respect to it ... For each word of the utterance that we are in process of understanding, we, as it were, lay down a set of our answering words. The greater their number and weight, the deeper and more substantial our understanding will be ... Any true understanding is dialogic in nature.
>
> (Voloshinov 1929: 102)

Thus, each student needs to 'lay down a set of answering words' to new ideas, in order to make them their own. A second purpose for interactive/dialogic talk is, therefore, to allow students to work with new ideas, and the teacher to be involved in supporting that process.

In the third cycle the teacher initiated a discussion of the results from the students' experiments in a whole-class format, identifying the main findings and also engaging the students in talking through how they might account for the anomalous result from one of the experiments. Here the purpose of the interactive/dialogic approach taken by the teacher was to make progress in developing the scientific story, not just by presenting the conclusions to be drawn from the experiments, but by attempting to involve the students in talking and thinking about the implications of the findings. The very act of conducting dialogic interactions in class serves to *model* this kind of intellectual engagement with scientific content. It helps to demonstrate to students that it is perfectly legitimate for them to 'talk science' in this kind of way (questioning and discussing findings and ideas, rather than just accepting them as being 'facts'). It helps to demonstrate to students that this is a productive and helpful way in which to 'think science'.

From this particular teaching sequence we have therefore been able to make the link from an interactive/dialogic communicative approach to three different teaching purposes: *exploring students' ideas*; *students working with new ideas*; *developing the scientific story*.

### Interactive/authoritative approach: 'work on'

While we acknowledge the fundamental importance of dialogic activity, it is also the case that in the classroom it is the *teacher* who has the responsibility for introducing and leading the development of the scientific story. A class of students could sit and *discuss* among themselves from now until doomsday, for example, the ways in which kinematics trolleys run down slopes and it is highly unlikely that they would ever stumble across the big ideas encapsulated in Newton's laws of motion. It is the job of the science teacher to intervene to introduce new ideas and terms, and to move the scientific story along. Authoritative interactions are an equally important and fundamental part of science teaching, and this, of course, is consistent with the fact that the social language of school science is, itself, authoritative in nature.

### Non-interactive/authoritative approach: 'review'

We also believe that there is an important place for those interventions where the teacher 'draws a line' under all the interactions (be they dialogic or authoritative) and makes a statement of 'where things are up to' in developing the scientific story and what is going to happen next.

Such interventions, to review and summarize progress, are often talked through by the teacher in a 'we' voice: 'What we've actually done is to try to . . . we've shown that . . . we can now see'. The implication of the 'we' is that there is a shared understanding (Edwards and Mercer 1987) within the class of the scientific story at that point. Of course, it is clear that this can never be the case, given the individual reconstructive step in learning that we referred to in Chapter 2. At the same time, we can see the importance of the teacher intervening to draw particular phases of activity to a close and punctuating the interactive staging of the scientific story with reviews of the 'current state of play'. In this way the teacher addresses the teaching purpose of *maintaining the development of the scientific story*.

It is with these ideas in mind that we draw attention to the value and importance of the *discuss, work on, review* rhythm. Furthermore, our experience of working in science classrooms has shown that it is *not* common to see examples of science teaching that move between all three elements in any sustained, systematic or rhythmic way.

### Staging the scientific story: a teaching 'spiral'

How do these ideas of the 'progressive transformation of content' and 'cycles of communicative approach' fit *together*? Figure 4.2 represents the way in which the staging of the scientific story proceeded for this particular teaching performance.

The movement between approaches is shown within each of the three cycles and, through these cycles of activity, the content of the classroom talk progressively shifts: from *places* to *possible things*; from *possible things* to *essential things*; from *essential things* to *scientific proof* for essential things. Putting the two trends together, we see the development of a *teaching spiral*, which emerges from the diversity of students' ideas and curls its way up towards the scientific point of view.

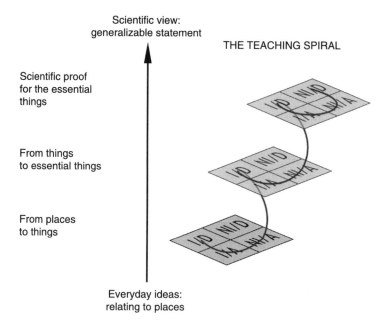

Scientific view:
generalizable statement

THE TEACHING SPIRAL

Scientific proof
for the essential
things

From things
to essential things

From places
to things

Everyday ideas:
relating to places

**Figure 4.2** The 'teaching spiral' for the rusting sequence.

What is the significance of this teaching spiral? First of all, we are *not* suggesting this as something to be aimed for in *all* teaching sequences. We do, however, strongly maintain (as argued above) that in *any* teaching sequence there *should* be variation in classes of communicative approach, covering both dialogic/authoritative and interactive/non-interactive dimensions. The rhythm of the teaching performance may not follow the elegant spiral seen here, but it should be consequent upon changes of approach.

A second point concerns the transformation of content and the nature of the *gap* between everyday and scientific views. In this particular case, both everyday and scientific perspectives are *empirical–descriptive* in nature, being based on observable features. As each of the teaching cycles is staged the teaching performance moves forward in a series of small, linked steps and the case study data demonstrate that, overall, the students were able to follow the performance successfully.

In other content areas of science, it might be that the teaching involves introducing a *theoretical explanation* (perhaps using the particulate theory of matter to explain certain physical properties of materials). Here the gap between everyday and scientific views is different, in that students are likely to have everyday ideas about the properties of materials in empirical–descriptive terms, and are now required to develop the theoretical–explanatory scientific view. The teaching performance might involve starting with empirical descriptions of various properties of materials, building on students' everyday knowledge, and then introducing the particle model to explain those properties. Here we imagine the path from everyday to scientific views as involving relatively small steps, with small *learning demands* (Leach and Scott 2002) through the initial empirical–descriptive phase, and then larger steps as the more demanding theoretical–explanatory model is introduced.

With this point in mind we now turn our attention to the next chapter, which focuses on just such a teaching sequence, where the aim is to develop theoretical explanations for various properties of matter, based on particle theory.

# 5  Struggling to come to terms with the scientific story

As a young boy I went with my uncle and aunt to visit the small farm of my aunt's father, in a remote country area of Minas Gerais state, in south-east Brazil. The spectacular images of man landing on the Moon were still very fresh in the minds of everyone, a powerful symbol of the 'global village' that the world was fast becoming through the influence of television. That people could sit in their living room in a small country town in Minas Gerais, miles from anywhere, and watch Neil Armstrong jumping like a child on the dry dust of the Moon's surface seemed like a miracle to me. In talking about the moon landing around the heavily laden dinner table I was shocked to discover that my aunt's father did not believe that it could be true. He argued forcibly that it was impossible to reach the Moon and that all the reports were nothing but a fabrication to deceive people. I was already a lover of science and tried all I could to convince him, but nothing would change his mind. Some days later, back in the noise and bustle of the big city where I still live, I read that some people had been arrested, accused of selling land on the Moon, using a Moon map to display the plots of land. I always think back to those times to remind myself that many people do not trust and understand science and many have little idea of the magnitude of scientific achievement.

We made the point in Chapter 2 that learning in the sciences can be a difficult and challenging process for students, especially when there is a big gap between *everyday* and *scientific* ways of talking and thinking about particular phenomena. With the rusting teaching of Chapter 4, this was probably not the case for most of the students. The everyday ideas that they came up with at the start of the sequence contained the seed of the scientific view, insofar as the part played by iron, water and air was raised in discussion from the first lesson. It is also important to recognize that the *learning demand* for these lessons was not unduly challenging, because the aim of the teaching was simply to establish the 'things needed for rusting'. In other words, drawing on the ideas from Chapter 3, the teaching was limited to developing an *empirical–descriptive* account of rusting, with no attempt to establish a theoretical explanation.

In this chapter we go into an area of science teaching and learning where the learning demands *are* high, and where the common-sense ways of everyday thinking can actually create *barriers* to understanding the scientific point of view. Here we focus on a sequence of lessons whose aim was to help students to understand the key features of the particulate theory of matter, and to see how it can account for the different states, and various physical properties, of matter. This teaching sequence takes us right into the realm of *theoretical entities*, which cannot be directly observed, and *explanatory accounts* of phenomena.

Of particular interest to us in analysing this sequence are the ways in which the *students* talk their way to an understanding of the scientific subject matter. We are interested in seeing the extent to which our framework, although developed initially with the teacher to the fore, can be usefully applied to capturing and characterizing the activity of students. With this in mind, the lessons we have chosen to focus upon include many instances of student–student talk, and we monitor and comment on those student discussions.

The intention here is not to provide a complete account and analysis of the whole teaching sequence, referring to each aspect of the framework throughout. Instead, we focus on particular episodes in the overall sequence, with a view to using the framework to identify, and reflect on, what we consider to be important and interesting features of the meaning making process.

## A teaching and learning performance

### The context

These lessons take us to a Brazilian Year 8 science classroom, in a school attached to the Federal University of Minas Gerais. Some of the students who attend the school come from a middle-class background, with many being the sons and daughters of the university lecturing staff. Others come from lower-middle-class families. Pedro, the science teacher who taught the lessons, is a skilful and well respected teacher with 12 years' experience in high school instruction. The particular class that we focus on consisted of 30 students, aged 14–15 years, including a wide range of student ability, with a significant skew towards the higher ability end.

The teaching sequence lasted for 12 one-hour lessons, extending over a period of three weeks, and focused on introducing the particle model to account for the different states, and various physical properties, of matter.

Pedro worked with a group of teachers and science education staff from the Federal University of Minas Gerais to develop the lesson sequence, as part of a professional development programme. He was therefore well aware of the impact of students' existing ideas on their learning of science content. The teaching sequence was planned in some detail, with each lesson including both whole-class and small-group discussion.

## The teaching approach

A typical approach to teaching the particulate theory of matter might consist of introducing particle models for solids, liquids and gases, referring to differences in particle separation and motion, and then using these models to account for various properties of matter. Pedro took a different course of action in starting with a series of demonstrations to show various properties of matter, and then working with the students to develop a particle model to account for those properties. The thinking behind this approach was that very often, when the particle model is presented authoritatively at the start of the teaching, the students do not have the opportunity to discuss its underlying key features. This can result in them developing alternative conceptions (see, for example, Driver *et al*. 1985) about both the particle theory and its applications that are not addressed. In the approach taken here these alternative interpretations were brought out into the open right from the beginning and explicitly addressed.

The lessons started with the students working in groups to develop 'models' for explaining various properties of matter. The teacher then focused attention on one of these phenomena, the compression of air in a sealed syringe, and the students worked to select and justify the model that they thought would best explain what they had seen. The students then presented their views to the whole class and the teacher used these ideas as a starting point for developing the school science particle model for *gases*. A similar sequence of events was followed in subsequent lessons for the other phenomena involving gases, and after this the teacher reviewed and summarized the features of a general model for *gases*, which he then related to changes in temperature and pressure.

At this mid-point in the sequence, attention was turned to explaining phenomena involving liquids and solids (including the expansion of mercury and the melting of naphthalene). This phase of the sequence concluded with small-group work and whole-class discussion, leading to the identification and summary of the key features of a particle model for solids, liquids and gases. The students were then given the opportunity to apply these models to other phenomena, focusing particularly on dissolving and diffusion.

Finally, the students returned to some work from the very start of the sequence, in which they had developed criteria for classifying solids, liquids and gases. Now the students were asked to review these criteria, this time drawing on their new understandings of the particle model. This led to an exercise in which the students discussed materials (including, for example, powders and foams) that appeared to resist classification as solid, liquid or gas. In the final two sessions the students worked on some written questions relating to particle theory and its application, and finally completed a formal test on the work.

### The teaching and learning episodes

We now turn to examining, in some detail, specific episodes from the lessons as the teaching and learning performance unfolds.

#### *Starting with the students' 'models'*

We first join the class with Pedro, the teacher, demonstrating various characteristic properties of solids, liquids and gases. For example, he used a sealed syringe to show that air can be compressed; a test tube, with a balloon over its neck, to show that air expands when heated; a mercury thermometer to demonstrate that the liquid mercury expands when warmed by hand. The teacher also referred to the smell of gas spreading in the kitchen, on escaping from its container, something the students might already have experienced, since kitchen gas (a mixture of butane and propane) is commonly used in Brazil. After the students had seen the demonstrations, Pedro distributed sheets of paper, each displaying drawings of the phenomenon, both 'before' and 'after' (for example, showing the sealed syringe, before and after compression). The students were then asked to draw or 'model' the solid, liquid or gas, before and after the transformation. As might be expected, a full range of ideas was produced by the students (see Figure 5.1).

Some of the students took an *empirical–descriptive* approach to the task and produced realistic, macroscopic drawings of the air inside the syringe. Others attempted to develop a *theoretical explanation* by using some kind of particle representation. At this stage it was difficult to ascertain the extent to which the students' representations map on to what might be considered to be an acceptable school science particle model. The particles shown in the students' drawings do not necessarily share the same features as those of the accepted scientific model. Indeed, it is difficult to see why they should, unless the students had already been exposed to this part of the school science social language. At the same time, however, it is important to recognize that, nowadays, particle (or 'atomic') notions of matter are very much part of our everyday world-view. It is therefore no great surprise that students should draw upon these ideas when asked to account for the various phenomena presented.

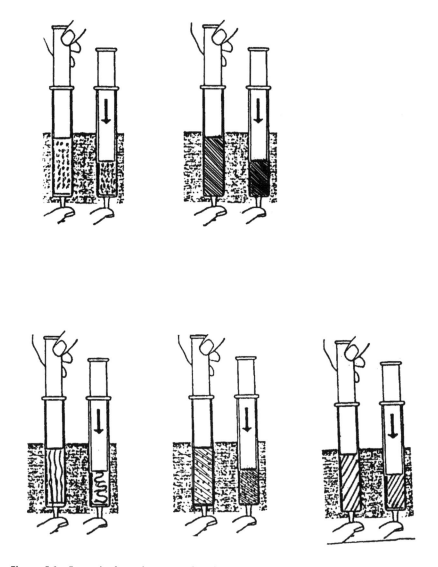

**Figure 5.1**  Examples from the range of students' 'models'.

Despite the variation in the kinds of representations produced, and the uncertainty about the students' thinking behind them, Pedro treated them all in the same way in framing the talk in terms of scientific models. From a teaching point of view the modelling activity served the dual purposes of *opening up the problem* and allowing the teacher to *explore the students' ideas* about these phenomena.

As the students worked on their representations it soon became apparent, from both the drawings and group discussion, that the question of 'what lies between the particles' was an issue for debate. From the scientific point of view, of course, there is nothing between the particles, but many of the students were showing air, or other substances, between them. Pedro recognized this and decided to address the issue by focusing the students' attention on the compression of air in the sealed syringe.

### Episode 1. Which one of these models, here, is the best?
At the start of lesson 2, Pedro had selected some of the students' models for compressing air in a sealed syringe, and drawn them on the blackboard. We join the class as the teacher is pointing to the drawings of the students' models on the blackboard.

1　Teacher: . . . OK, I want you to discuss these models in groups. Which one of these models here [counts the models on the blackboard] is the best, OK? I mean, the one that best explains the compression of air in a sealed syringe. In deciding this, you should take the following into account. The best model has to take into account . . . the conservation of mass . . . Everyone has agreed that mass is conserved and that density varies. Haven't we? Now, one more thing . . . what I want to know is . . . the air, is it continuous [as suggested by some students] or discontinuous [as suggested by other students]? What does it mean for the air to be continuous or discontinuous? If the air is continuous, it means that it has no particles, doesn't it? If the air is discontinuous it has particles. Now . . . it might also be like he [a student] said [points to the second model], you can think that air is a mixture, something continuous and discontinuous, like here [points to the second model]. Now, another thing. If the air is discontinuous, what is there between the particles? In other words, this space here, what is this? [points to the first model] OK? You should try to answer this as well. So, you are going to discuss which model is the best, based on these ideas . . .

This is an interesting statement from Pedro. As he briefs the students about the task, he reviews the various 'models' (both discontinuous and continuous) shown on the board, adopting a *non-interactive/dialogic* communicative approach. In addition, Pedro alerts the students to the question of what lies between the particles, and suggests that they 'should try to answer this as well'. In this way, Pedro not only invites the students to discuss the different models at hand, but also attempts to retain some influence on those discussions. Bakhtin (1934) makes the point that talk can be considered to be dialogic not only in terms of representing different points of view, but also in being oriented towards the likely response of the listener. Here, Pedro shapes his initial statement in a dialogic manner, by not only representing the students' points of view but also anticipating their responses. We now turn our attention to one of the groups of students as they discuss the various 'models' on the board.

### Episode 2. Air is all around, not only in the dots

As we join the group, Carolina one of the girls, is seeking the views of the rest of the students:

1 Carolina: Which do you think is the best?
2 Edward: That one with dots. [Edward is referring to the dots in the drawing]
3 Carolina: I think the most scientific one is the one with dots.
4 Raquel: But, that one with dots . . . what's that empty space? Cos air is all around, not only in the dots. So, we could colour all around . . . that's what I think. Did you hear what he said? [referring to the teacher] We should think about the space . . .
5 Carolina: But then it's not that one with the dots . . . it's that one. [she points to another drawing]
6 Raquel: I think we could colour it like this. You make this first one very light . . . then, when you compress it . . . it would be darker because it's more concentrated. Then there won't be any space and there won't be any doubt.
7 Carolina: But air isn't continuous like that . . . air is made up from lots of particles.
8 Raquel: [Asking Edward] What d'you think?
9 Edward: . . . the air . . . is . . . like she said, air has particles.
10 Raquel: But, between the particles, is there a vacuum?
11 Carolina: No, but . . . here we don't have only air particles, we have other particles . . . nitrogen, pollution, dust . . . all sorts of things.
12 Raquel: Hum . . .

So the students take an *interactive/dialogic* approach to try to reach agreement in a situation where two different views (the 'dot' model and the 'continuous' model), representing two different ways of conceptualizing matter, interact. Pedro's guidance in the initial briefing, about the space between the particles, proves to be influential in the development of the discussion. Thus, when Raquel (turn 4) refers to the space between particles – 'We should think about the space' – it is not her voice that is speaking, but the teacher's, and this contributes to framing the students' negotiation of meaning.

Raquel would prefer to use a continuous representation, because then 'there won't be space and there won't be any doubt'. It is clear that her thinking is guided by the notion that air fills all of space, as she comments, 'air is all around, not only in the dots'. This is an articulation of the commonly held everyday view that 'nature abhors a vacuum', a belief about the very nature of the natural world (referred to as an *ontological commitment* in Chapter 2) that is likely to be a significant and influential part of the students' thinking. Pedro's question thus prompted the students to make explicit, and to address, this view that underlies their thinking.

Although Raquel's view is not accepted, it demands consideration by her classmates, and the model that emerges at the end of the episode reflects a kind of compromise. Thus, Carolina accepts the counter-argument from Raquel ('But, between the particles, is there a vacuum?') and refines her particle model so that the empty space is filled with 'Nitrogen, pollution, dust . . . all sorts of things'. Carolina's list of 'particles' appears to include a mixture of theoretical (nitrogen particles) and everyday (dust particles) entities.

After all the groups have completed their discussions, Pedro collects their ideas and draws them out on the board, helping the students to make their points of view clear, but not interfering with the content.

### Episode 3. Let's imagine a flask full of marbles

It is clear from the students' drawings that the question of what lies between the particles is a big issue for the whole class, and not just for Carolina's group. Pedro attempts to address this issue by introducing an analogy in which he likens a flask full of air to a flask full of marbles.

1 Teacher: . . . If the air was made up from particles, which can be shown as marbles, the air itself would be the marbles. So, I mean . . . so, what is there between the marbles?

2 Students: [in chorus] Air! . . . Other gases!

3 Teacher: No. What I'm saying is that the marbles *themselves* represent the particles of air, ok?

4 Julia: Between them there are more particles.

5 Teacher: No! Let's imagine a flask *full* of marbles. Is there any way of putting more marbles inside the flask? It's already *completely* full, but there is still some empty space between the marbles, isn't there?

6 Anita: But . . . we think that the air particles are different from the marbles.

7 Teacher: What *are* the particles of air then?

8 Students: They're the marbles.

9 Carolina: Ah! I get what you mean . . .

10 Student: A ball of plastic foam . . . [one of the students calls out]

11 Raquel: Yes . . . of plastic foam!

12 Carolina: No! No, it's all the same!

13 Raquel: No, because when you squeeze them they will . . . [makes a hand gesture of compressing]

14 Students: Unintelligible. [several students speaking excitedly at the same time]

15 Teacher: Look, even with the plastic foam, there would still be some space [between the particles], wouldn't there?

16 Erica: But can the particles have another shape?

17 Teacher: What shape?

18 Erica: Square.

19 Teacher: Oh! So, here is another model! What is it? Pay attention! The model that Erica is suggesting is that the particles, little cubes, all fit perfectly together. Is that right?

20 Erica: Yes.

21 Teacher: This is a new model. Pay attention! Let's draw it on the board. [he draws the model on the board]

22 Teacher: But then, what happens? When I try to compress this thing here [pointing to the drawing], there is no space at all, OK? Will I be able to compress it?

23 Students: [in chorus] No!

24 Teacher: Only if what?

25 Carolina: Only if the particles are elastic . . .

26 Teacher: Only if the particles are elastic, OK?

The students become very excited as the episode develops and the majority are fully engaged in following the debate, whether contributing directly or not. It is clear from the exchanges that it is the students' commitment to the notion that 'nature abhors a vacuum' that underlies their thinking and leads them to adjust their model, and even invent another one, to counter what Pedro is saying. At the end of the episode the model that has emerged is a *substantialist* one, where the cubic particles of air fit perfectly together, and can be compressed themselves.

It is highly unlikely that the students are fundamentally committed to a view of matter based on 'foam cubes'. This 'model' has evolved as a way for the students to avoid the idea that there may be nothing between the particles. Pedro now faces these deep-seated beliefs head on, working in the gap between everyday and scientific views, and it is not difficult to imagine the teaching challenge (and possible discomfort!) that he is experiencing. Both empirical and analogical evidence have proven to be insufficient in challenging those beliefs, and have actually been drawn upon by the students to reinforce them.

The episode as a whole is *interactive* and highly *dialogic* in character. Right at the start Pedro (turn 1) attempts to develop an authoritative I–R–E interaction to establish that there is nothing between the particles. After Anita's intervention (turn 6) the pattern of discourse changes completely and we have seven turns (8–14) of speech where students talk about the particles, without any interference from the teacher. When they come to the conclusion that the particles might be like compressible plastic foam, Pedro returns to the issue (turn 15) of the spaces between them. Erica then *initiates* another sequence of interactions by suggesting (turn 16) an alternative shape for the particles. The teacher accepts this, enacting a confirmatory exchange with Erica (turns 19–21), and then establishes through an I–R–E pattern (turns 22–26) that this model will work 'only if the particles are elastic'.

### Episode 4. And what is this space? It's empty space!

At this point, Pedro confronts the class with a basic question:

Teacher:  So, what's the problem with the particle model?

The class responds as one:

Students:  [in chorus] It's the empty space!

Pedro asks why they are not willing to accept the idea of empty spaces between the particles and various students reply, 'We don't think there is an empty space.' Pedro now returns to reviewing some of his earlier arguments about the empty space, but this time he presents them in an *authoritative* way.

1  Teacher: OK, for the air to move . . . if the particles are moving . . . for the movement to exist what else must there be?
2  Students: [in chorus] Empty space!
3  Teacher: There must be empty space, OK? For the particles to move space must exist. And what is this space? It's empty space! The condition for the existence of movement is that there must be empty space. I mean, if there was no space the particles couldn't move, OK? We know that the particles move, we are going to do an experiment about this. But if the particles can move, there must be space between them . . . and this space is empty.

Pedro also emphasizes the importance of having a model that does not depend upon the nature of the constituent particles.

4  Teacher: I can use that model [referring to the particle model] without having to talk about the exact nature of the particles. I don't need to say whether the particle is a ball or a square, whether it is elastic or non-elastic. It's better to work with a model that doesn't specify the nature of the particle, since I can't observe it. We can't attribute to the particles all the properties we see in the substance.

So Pedro presents the particle model for gases, moving from an *interactive/ authoritative* approach, played out in an I–R–E pattern (turns 1–3), to a *non-interactive/authoritative* presentation. In turn 3 Pedro *repeats* the idea of 'empty space' seven times in 20 seconds, marking it out as a key idea in no uncertain terms.

### The opening phase: the rhythm to the discourse

What can be said about the way in which the classroom talk has developed over this first phase of the teaching sequence? Are there any similarities with the rusting teaching of Chapter 4?

The sequence started with Pedro demonstrating various physical phenomena, and the students working in groups to 'model' what happened with each. Pedro then directed the groups to focus their attention on the compression of air. So, the teacher *opened up the problem*, encouraged the students to discuss their ideas in an *interactive/dialogic* way and gave himself the opportunity to *explore the students' ideas* (identifying the issue of what lies in the space between the particles). The approach is similar to that taken by Lynne, as she set up the nails display and discussed possible 'things needed for rusting' with the students.

At this very early stage, however, rather different paths emerge for the two sequences. In the rusting lessons, the ideas offered by the students were largely consistent with the school science point of view (including the 'key things' water and air), and so provided the seeds for the development of that view. Here, the models developed by the students included features that are fundamentally at odds with the school science view (that 'stuff' should exist between the particles). These features appeared to follow from a basic commitment, by the students, to the idea that 'nature abhors a vacuum' and actually presented a significant barrier to development of the school science view.

So, whereas Lynne was able to work on the students' ideas in a relatively unproblematic way, Pedro needed to make a much stronger intervention (using the marbles analogy) to challenge the students' views and to develop the scientific model. As it happens the analogy did not prove to be effective (as the students subverted it to their own way of thinking) and this phase of the teaching ended with Pedro authoritatively setting out the school science particle model for gases. Lynne was able to make an authoritative presentation to summarize what seemed to be a largely 'shared view', with teacher and students on convergent paths. Pedro, on the other hand, was summarizing the scientific view as an alternative perspective introduced by himself, and one that differed from, and conflicted with, the students' views. Although both teaching sequences opened with a *discuss, work on, review* pattern, it is clear that these steps have quite different functions in each sequence, and this is due to the differing relationship between everyday and school science knowledge in the two cases.

### Episode 5. What happened is the particles got bigger

Pedro now returns to the demonstration in which he heated some air in a test tube, so that it expanded and filled a balloon. Having drawn on the blackboard some of the students' 'models' for the air, before and after heating, he asks the groups to select one that would best explain the air expanding. At this point, we return to our group of students.

1   Carolina: What happened is the particles got bigger.
2   Raquel: The particles expanded . . .
3   Carolina: Expanded . . .
4   Raquel: That's model three. [referring to the number of the model on the sheet distributed by the teacher]
5   Carolina: What do you think, Ricardo?
6   Ricardo: Nothing . . . I don't know.
7   Carolina: Hey! We have to answer here. We saw the balloon fill up, didn't we? But we have to answer . . . explain . . .
8   Raquel: We have to explain, air, when heated, expands.
9   Carolina: Expands. The air particles expand when heated because there's empty space between the particles.
10  Edward: It's the air that expands.
11  Carolina: It isn't the air that expands! It's the particles that expand.
12  Raquel: Ah, now we have to draw here . . . the model that we've chosen . . . [they start drawing]
13  Carolina: Here, look, we have to describe the model. How do we describe the first [test tube]?
14  Edward: Normal.
15  Carolina: Normal. The particles have their normal size. Now, in the second [test tube], they've got bigger, expanded, filling a bigger volume, haven't they?
16  Edward: Yes.

Although Pedro has just made his authoritative presentation of the particle model for gases, emphasizing that 'we can't attribute to the particles all the properties we see in the substance', the students select a substantialist model, where the particles themselves expand. From the outset, Raquel and Carolina agree that the particles expand and in doing so develop what appears more like an *empirical description* than a theoretical explanation. Ricardo and Edward offer little by way of challenge to their view, and so the interactions become predominantly *authoritative* in nature. In contrast to the discussion of the compression of air, where 'counter-words' had an important role to play in building a consensus, here there is room only for the substantialist point of view. Carolina's emphatic statement (turn 11) in responding to Edward's objection that 'It's the air that expands' suggests that she was not prepared to negotiate her point of view.

While the group has once more taken a path that diverges from the scientific perspective, the idea of 'empty space between particles' no longer seems to be a problem. Indeed, in turn 9 Carolina uses this concept to explain how air particles are able to expand: 'because there's empty space between the particles'.

Perhaps the most striking feature of this episode relates to the communicative approach taken. Although Pedro's point of view is not represented here (as with the compression of air in Episode 2), the teacher's role is assumed by Carolina. As a consequence, the episode has a clear interactive/authoritative nature, where the authority comes neither from the institutional position of the teacher (who was not present) nor from the authority of the scientific view. Instead, the authority emanates from the leadership of Carolina and Raquel in the group. Apart from admonishing a recalcitrant Ricardo – 'Hey! We have to answer here' – (turn 7), Carolina makes an intervention to review the 'particles expanding' view through a very obvious I–R–E, confirmatory exchange with Edward.

First, Carolina asks 'How do we describe the first [test tube]?' (initiation) and Edward replies, 'Normal' (response). This is evaluated by Carolina: 'Normal. The particles have their normal size' (evaluation). She then starts a new triadic interaction by asking a further question: 'Now, in the second [test tube], they've got bigger, expanded, filling a bigger volume, haven't they?' (initiation). Given the way that Carolina frames the question, there is really only one possible response and Edward duly provides it – 'Yes' (response) – thereby completing the confirmatory exchange.

An important observation from this episode is that classroom talk can be authoritative or dialogic in nature, irrespective of whether it is led by a teacher in whole-class discussion or by a student working in a small group. The episode also demonstrates that students can take on different roles in the classroom, including that of *teacher*. The asymmetry between the teacher's and the students' roles, which is reproduced in this interaction between students, seems to be an inherent sociocultural and institutional characteristic of schools that frames the discourse, even when led by students in the absence of a teacher.

When the individual groups have completed their discussions, Pedro invites them to report back. In the whole-class talk that follows, Pedro is able to explore the students' views, dealing with the 'particles expanding' problem as it arises. In contrast to the discussion of 'space between the particles', these interactions are a little more straightforward for the teacher to handle. For example, when Carolina reports that her group has chosen the 'particles expanding' model, Fabiola, one of the other girls, laughs and calls out:

Fabiola: Expanding involves increasing the space between the particles. The particles themselves don't expand, Carolina. They can't do that!

Once again we see a student, in this case Fabiola, taking on the role of *teacher* in evaluating a fellow student's point of view. This kind of intervention certainly made Pedro's task of addressing the substantialist obstacle easier, since some of the class already shared the scientific view and were able to pressure other students to adopt it. By this point in the lesson sequence, the problem of 'empty space between the particles' seems to have been overcome with the class as a whole, as it made no appearance in the discussion.

### Episode 6. It's nothing to do with that! Forget the butane bottle!

We now move on to the fourth lesson of the sequence. By this point, the students have discussed each of the initial demonstrations in terms of a particle model, and Pedro now asks them to 'draw a particle model for all gases'. This task signals a move away from using the particle model to provide *theoretical explanations* for specific phenomena, to presenting a theoretical model that represents all gases. In other words, the students are being asked to present a *theoretical descriptive generalization* for gases. 'Gas' is now being considered as an abstract entity, a general category of matter, rather than as a real and concrete substance. We join the group with Edward repeating the teacher's instructions.

1  Edward: It has to be a model for all gases . . . [parallel talk, inaudible]
2  Carolina: What does a model for gas have to have? What characteristics does it have to have?
3  Edward: First, being gaseous.

This initial exchange between Carolina and Edward suggests that they understand what Pedro has asked them to do. However, Alex now introduces the idea of a 'container' for the gas.

4  Alex: No, [it has to] hold [i.e. contain] the gas.
5  Carolina: Yes, to hold the gas.
6  Alex: It has to be compressed, of course. You can't put a gas . . . no, because the gas in a butane bottle [botijão].
7  Carolina: . . . is liquid.
8  Alex: It's liquid, but . . .
9  Edward: For instance, you can't put [it] in a plastic container. Of course, it has to be something . . .
10 Alex: But, you can't put the gas . . . how can you transform it into its liquid state?

11   Ricardo: Put it in a container . . . [inaudible]
12   Carolina: Hey folks! Then it must be a butane bottle.
13   Edward: What did you say? . . . A container . . . I forget the word . . . it has to be a container. [Edward gesticulates with his hands, indicating something closed]
14   Carolina: Gee! Seriously, I have no idea.

This seems very confused on first reading. In fact, Alex is thinking about a specific gas, the cooking gas that is widely used in kitchens in Brazil (a botijão is the metal container for such cooking gas), and the teacher's reference to 'gases' (in general) seems to have sparked this association. In addition, Alex's interest in a container for the gas may have been prompted by Pedro's instruction to the students to draw the gas 'in a box' or 'square'. Whatever gave rise to Alex's ideas, his intervention succeeds in completely redirecting the talk of the group and they proceed to discuss butane cooking gas, and how it might be contained. When they start to struggle with these ideas, they call for the help of the teacher, who is talking with some other students.

15   Alex [to the teacher, who comes over to the group]: Sir, can all gases be transformed into a liquid state?
16   Teacher: Yes, all can be transformed, but some have to be in perfect condition, otherwise they might blow up.
17   Alex: Such as . . . ?
18   Teacher: Like methane, natural gas. What do they do with natural gas? [Pedro goes on to explain why natural gas must be transported through pipelines and cannot be stored in containers like butane and propane]
19   Edward: Look! The problem has changed completely . . . do we put it [the gas] in a pipeline or inside a butane bottle?
20   Teacher: Hey! Listen! What are you meant to be doing with this question? You have suggested a model for each of the previous situations. This model now, the final one, isn't it the same model for all situations? So what is it? You need to generalize. What *is* that general model?
21   Alex: It's a compressed model, isn't it?
22   Teacher: No! It's nothing to do with that! Forget the butane bottle! It's made up of . . . ?
23   Carolina: Particles . . .
24   Teacher: . . . of particles, yes! All the gases, then . . .
25   Alex: Particles!
26   Teacher: Particles. And what are the characteristics of the particles? How the particles . . . what happens to them . . . how do they behave? [Pedro leaves to join another group]

After responding to Alex's question about liquefying gases, Pedro realizes that the group are hopelessly off-task, given the aim of the exercise. He advises them in no uncertain terms to 'forget the butane bottle!' and the group returns to talking about 'gases'.

27 Alex: The thing is, it has to be made a characteristic of a gas, of gases . . .
28 Edward: We were talking about that, totally wrong! This doesn't have anything to do with a container . . .
29 Carolina: But it's about describing the model.
30 Edward: Maybe he said, 'draw a silly box and put the model inside the box'.
31 Carolina: Yeah, that's it! [they begin to draw] Are the particles of a gas the same as the particles of air?

The group seems to be back on task now, although Carolina's final question (turn 31) suggests that there may still be doubts about the nature of the question they are addressing. It is interesting to note that this episode occurs at the precise point in the overall teaching sequence when Pedro attempts to move the scientific story towards a theoretical scientific discourse involving abstract categorization. The intellectual task of leaving behind explanations for specific empirical phenomena, and thinking about gases as a general category of matter, certainly proves to be a significant one for these students. At the heart of it is coming to understand just what is meant, in the scientific social language, by a *theoretical descriptive generalization* for gases.

### Episode 7. The greater the energy, the faster the motion, and that's it!
For the next episode we skip ahead to the tenth lesson, near to the end of the teaching sequence. We join the students as they are engaged in various activities aimed at reviewing and summarizing the characteristics of the particle model. Here the group is addressing the question: 'How is energy related to particle motion?'

1 Raquel: So, how is energy related to the particles' motion?
2 Edward: We think that it's the gas, it has more motion, it spreads out into the surroundings.
3 Raquel: And so?
4 Edward: Taking account of the movement of the particles, when they have more motion it's in the gaseous [state]. The one with more energy is the gas.

At the outset Edward seems to understand the question, as he tries to relate the energy to motion in the gaseous state.

5 Alex: Look at how my three fit together nicely. [Alex is referring to another question]

6 Raquel: But now, how is energy related to the particle motion?

7 Alex: We only have to take into account ... as a gas has more motion, it has more energy. Now, ... the problem is how to explain a gas.

8 Raquel: In the gaseous state the particles have more energy and their motion is greater ... wait a minute ...

9 Alex: Please, you're driving us crazy!

10 Raquel: Yes, I'm crazy too!

Alex is distracted by a previous question, but Raquel brings him back to the issue under consideration. It becomes clear (turns 9, 10) that both Alex and Raquel are deeply engaged in trying to think the question through, and then Carolina intervenes to broaden the discussion by referring to solids and liquids:

11 Carolina: If we were talking about solids, how much energy is associated with the motion of those particles? Because it's solid and they vibrate, I mean ... then ... if we choose the liquid it means that they move themselves in groups, but if we choose the gas they move with more freedom.

12 Raquel: But it has to be logical.

13 Alex: Yes, you can't just say, 'it is because it is', 'it is because it is' ...

14 Carolina: But it *is* because it is!

15 Ricardo: The gas ... when a gas is heated, its volume increases and its motion increases as well.

16 Edward: But you can't say that a gas has motion only when it's heated.

17 Raquel: Why is its motion greater than that of solids and liquids? Because the particles move individually.

18 Edward: Because they can be considered as individual particles.

19 Carolina: So?

20 Raquel: So, they can move faster.

21 Edward: Yes! Move faster.

22 Raquel: Then we have to explain it this way. We can say that because they move themselves ... no, they move themselves because of what?

23 Edward: Because they are individual, the particles are individual ... it's that.

24 Alex: I'd say that they're independent and move more quickly, then ...

25  Raquel:  Yes, you need to say, they move more quickly because . . .
26  Alex:  How is energy related to motion? How is the amount of energy
related to motion? The greater the energy, the faster the motion, and
that's it!

The students develop a discussion around the concepts of particle energy,
motion and arrangement. In doing so they review several characteristics of
the particle model for solids, liquids and gases, including: the characteristic
particle motion for each state; how heating affects the volume and particle
motion in gases; the intrinsic particle motion in gases ('you can't say that a gas
has motion only when it is heated'); that the particles in a gas are far apart and
therefore 'can be considered as individual particles'; and that being individual
they move faster.

Some of these statements are not strictly accurate. For example, the
particles of a gas do not necessarily move more quickly than the particles of
a liquid, since kinetic energy (and consequently the motion) depends on
the temperature and not on the particles' condition of being 'individual' or
not. Nevertheless, the students are addressing the question with much more
appropriate variables (motion, energy and arrangement) than in previous
episodes. The teacher's voice is present throughout the episode, but in a
way that is different from the earlier episode, where it was referred to by the
students as a point of guidance: 'Did you hear what he said? We should think
about the space.' In this episode, there is strong evidence to suggest that the
students have been able to take on, or appropriate, the teacher's voice as they
are now able to articulate many features of the school science story. It is no
longer a case of the teacher's voice speaking through the students. There is
evidence of internalization having taken place, with the students now being
able to use the scientific view as their own.

Here the students take an *interactive/dialogic* communicative approach.
Although the students' statements draw only on the school science point
of view, they are used as a basis for generating new meanings, in addressing
the question 'how is energy related to the particle motion?' The students use
logical prompts ('and so?') to maintain the continuity of thinking throughout
the episode, and there is a strong impression of new ideas being developed on
the individual plane, as a consequence of the talk on the social plane.
Although Raquel takes on the role of teacher briefly at the start of the
episode – 'But now, how is energy related to the particle motion?' (turn 6) –
the episode as a whole shows no triadic patterns of interaction. Most of the
statements build on previous ones, generating chains of interactions, of an
I–R–F–R–F– form. For example:

17  Raquel:  Why is its motion greater than that of solids and liquids?
Because the particles move individually.

18   Edward: Because they can be considered as individual particles.
19   Carolina: So?
20   Raquel: So, they can move faster.
21   Edward: Yes! Move faster.
22   Raquel: Then we have to explain it in this way. Then we can say that because they move themselves . . . no, they move themselves because of what?
23   Edward: Because they are individual, the particles are individual . . . it's that.

Raquel starts by posing a rhetorical question (initiation) and Edward repeats Raquel's answer (response). Carolina provides a prompt to maintain the development of the argument (feedback). Raquel answers again (response) and Edward responds positively (feedback). Raquel then poses another question (initiation), Edward answers (response) and so the dialogue continues.

During the episode the students use theoretical generalizations about solids, liquids and gases fluently, the uncertainty of the previous episode has apparently disappeared and the students are quite happy to talk in terms of general categories of matter. The question 'how is energy related to the particle motion?' is an interesting one, in that it takes the students to yet a further level of abstraction or generalization. The focus is no longer on particle motion in a particular state of matter (solid, liquid or gas), but now involves seeking a relationship between energy and particle motion *across* the states. The group starts by focusing on the gaseous state, widens the discussion to consider solids and liquids and then finally Alex makes his triumphant statement: 'How is the amount of energy related to motion? The greater the energy, the faster the motion, and that's it!' There is a strong impression here of Alex suddenly coming to understand the nature of the question, and as he does so, he is absolutely certain of the validity of his answer.

A further interesting feature of this episode is the way in which the students start to make statements about the *nature* of scientific knowledge:

12   Raquel: But it has to be logical.
13   Alex: Yes, you can't just say, 'it is because it is', 'it is because it is' . . .
14   Carolina: But it *is* because it is!

What the students are talking about here relates to *epistemological* features of the scientific story, making judgements about statements that must be logical and non-tautological to be scientific.

### Episode 8. I can do it . . . but it doesn't go inside my head!

Right at the start of the teaching sequence, the students were asked to classify a collection of different materials as solids, liquids or gases, and to explain the criteria used to inform the classification. The outcomes of this activity were put to one side and returned to in this final activity of the sequence.

Pedro grouped the criteria developed by the students in the initial exercise, and summarized them for the students in three basic classes:

1   *Perceptual/sensory*: proposed by students at the start of the lessons. For example, a solid is hard, we can hold it; a liquid is wet, we cannot hold it.
2   *Empirical*: also proposed by the students in the initial activity (having been addressed in previous science classes). For example, solids have constant volume and definite shape, liquids have constant volume but take the shape of a container.
3   *Particle model*: drawing on the ideas developed in these lessons.

The students were then asked to review these criteria and to apply them to some materials that appeared to *resist* classification. Here we join the group as they are talking about glass:

1   Raquel: Look at 'glass', an amorphous material, it doesn't have a geometrical arrangement of particles, its particles are disorganized. If the particles are disorganized, it's a liquid.
2   Edward: If you say that we can hold it, then it's a solid . . .
3   Raquel: There should be only one classification, only one criterion, otherwise there is no way . . .
4   Alex: Just because it's solid doesn't mean that the particles are organized.
5   Raquel: Then they are disorganized, but really it's a solid.
6   Edward: It depends on the criterion you use for solids.
7   Raquel: Even at a glance, you can see that it's a solid.
8   Alex: Look, just between you and me [Alex lowers his voice to avoid being recorded], I think that glass *is* a solid, if I'm in the street, there's no way I would use particles to define it.
9   Carolina: Neither would I.
10   Raquel: Solid outside. Inside the laboratory I use particles.
11   Alex: Because he [referring to the teacher] asks for it. Now, it's a liquid when it makes something wet.
12   Carolina: Hold on folks! Glass . . . it's useless to say it's a liquid, it's a solid, it's an exception to the rule, all rules have exceptions.
13   Edward: You can hold solids.
14   Carolina: No, no, the ability to think that glass is a liquid. I mean . . . I can do it because I know the particles are disorganized, but it doesn't go inside my head!

The students thus bring together two social languages, their newly acquired school science view and an everyday way of talking and thinking. The two forms are brought out into the open from the beginning as Raquel focuses attention on the example of 'glass' and declares that it must be a *liquid* because the particles are disorganized. Edward offers an alternative point of view, by drawing on an everyday perceptual criterion, and argues that it is a *solid* because you can hold it. The existence of the opposing points of view creates some consternation, and Edward arrives at the conclusion that 'it depends on the criterion you use for solids'.

Of course, Edward is absolutely correct in his observation and the point takes us back to Chapter 2, where we considered the question of what is involved in learning science. There we argued that learning science involves developing an alternative way of talking and thinking about the natural world. It involves learning the social language of school science and adding it to a personal 'toolkit' of ways of talking and knowing about the world, which can be drawn upon as appropriate, in different contexts. If the students accepted such a view of learning, it might allow them to accept more easily the possibility of identifying glass as a 'solid' in everyday situations, and as a 'liquid' in scientific contexts.

It is clear, however, from the secretive style of Alex's whispers (turn 8) that he, and the others, believe that science is expected to provide the single truth about the way things are, and this is what they must accept. They are not happy with the multiplicity of interpretations that follow from using different social languages: 'if I'm in the street, there's no way I would use particles to define it'. Raquel and Alex agree that they will use particles 'in the laboratory', but only because 'he asks for it'. Ironically, these 'subversive' remarks from the students are consistent with the 'toolkit' notion of learning. Here is a teaching and learning episode where it seems that the students' learning of science would be helped if they were provided with some insights into what is involved in learning science.

It is tempting to suggest that the epistemological hiatus that has arisen in this episode is due to the fact that glass is unusual in being a 'solid' material with an amorphous structure. In fact, the majority of solids that we deal with in everyday life do not have regular geometric arrays of particles. As outlined in Chapter 2, defining solids as having regular particle structures is very much a perspective of the school science social language, and one that differs from both everyday and scientific views.

The outcome of all of this is that the students *understand* the scientific point of view, but they have major problems in *accepting* it. The scientific view is *intelligible* to the students, but not at all *plausible* (Posner *et al.* 1982). Here we are reminded of the words of another student, this time English, after some lessons on the particulate theory of matter.

Well, the thing is say with God. People say there is a God, and all that, but it's pretty hard to actually believe there is one. And I think it's the same with atoms and stuff like that. There must be atoms making things up but they're pretty hard to believe.

(Words of a 14-year-old after lessons on the particulate theory of matter)

## Coming to terms with the scientific point of view

In analysing these lessons, we have focused on one group of students and developed a detailed account of the way in which they progressed in coming to terms with the school science point of view. We now wish to step back a little, to take an overview of the way in which this was achieved.

### A series of barriers to learning

One striking feature of these lessons is the way in which the students were confronted by different problems, or 'learning barriers', to be overcome as they tried to develop an understanding of the school science point of view. Looking back over the analysis, the main barriers can be summarized as follows.

#### Barrier 1: 'nature abhors a vacuum'
A fundamental feature of the particle model of matter is that there is *nothing* in the space between the particles. A significant problem therefore arose for the students, as they started from an everyday way of thinking, where 'air is everywhere' and 'nature abhors a vacuum', and moved towards accepting that there is nothing between the particles in gases, liquids and solids. Here, the learning demand is essentially *ontological* in nature, as the students are required to review and change a basic commitment about the nature of the world.

#### Barrier 2: 'a substantialist view of particles'
A further basic feature of the particle model is that the particles themselves are immutable, and that the physical properties of a substance are explained not in terms of changes to the actual particles themselves, but in terms of changes to their motion and distribution. In one of the lessons, the students explained the expansion of air not in terms of the increased separation of the particles on heating, but as being due to the expansion of the air particles themselves. The problem here is one of moving beyond empirical descriptions of matter, where it is assumed that the expansion of the substance is due to the expansion of its constituent parts. The student must learn to conceptualize

matter in terms of unchanging particles and come to appreciate that this single model can be used to explain a whole range of substance properties. The learning demand is therefore both *conceptual* and *epistemological* in nature.

### Barrier 3: 'leaving behind specific phenomena'
This problem occurred as the students were asked to move beyond developing particle explanations for specific phenomena, such as the expansion of air, to consider a particle model for *all* gases. Here the learning demand is largely *epistemological* in nature, as the students are required to think about 'gases' as a general category of matter, and to develop a generalized theoretical description for that category.

### Barrier 4: 'leaving behind specific states of matter'
This final barrier came into operation as the students were asked to develop a relationship between two properties of particles (energy and motion) that applies across all three states (solids, liquids and gases). As we saw in the account of the lesson, the greatest hurdle for the students here seemed to involve understanding the nature of the question. The learning demand is *conceptual/epistemological* in nature, as the students are required to move from their knowledge of the behaviour of particles in the individual states, to develop a relationship that applies across the states.

These barriers appeared as the scientific story took the students on a path that moved further and further away from the actual phenomena demonstrated by the teacher in the first lesson. In this sense the teaching sequence involved the same kind of progressive recontextualization as that identified for the rusting teaching. In the case of the rusting teaching, however, the content remained focused on an *empirical–descriptive* account, while here it is clear that there was a much greater shift towards the abstract:

- *from* empirical descriptions of specific physical phenomena;
- *to* theoretical explanations for specific phenomena;
- *to* theoretical descriptive generalizations for specific states;
- *to* theoretical descriptive relationships across states.

Viewed in this way, it is hardly surprising that the development of the scientific story generated significant learning demands for the students, and that as the story went deeper into the social language of science, conceptual, ontological and epistemological challenges emerged.

An interesting feature of these lessons was the way in which the students were able to overcome each of the learning barriers, take on the scientific point of view and then use it as the accepted way of dealing with problems in subsequent tasks. What was once a major barrier to progress became a tool for dealing with the next problem. For example, the concept of empty space between particles was initially met with a lot of resistance by the students, but just after the teacher's intervention the students were fluent in using the idea to justify why 'air particles expand when heated'. In a similar way, although the students initially struggled with the idea of a general particle model for 'gases', they soon showed themselves to be quite comfortable in making statements about the particle models for solids, liquids and gases as general classes of matter.

### The teaching approach: exposing everyday views

We made the point earlier, in relation to the opening phase of the sequence, that in these lessons the teacher adopted a similar cyclical pattern of communicative approach to that identified in the rusting lessons:

- 'discuss' (interactive/dialogic);
- 'work on' (interactive/authoritative);
- 'review' (non-interactive/authoritative).

The big difference from the rusting lessons was that the outcome of the initial discussion included some fundamental ideas that were very much at odds with the scientific point of view (such as there being no empty space between particles). The teacher therefore needed to make much more authoritative interventions in trying to challenge the students' everyday views, and to develop the scientific perspective. Indeed, one might question the pedagogical wisdom of asking the students for their ideas about why matter behaves as it does when it is inevitable that they will respond with these kinds of everyday views. This question does not apply solely to teaching about matter, it is a fundamental issue for the whole of the science curriculum, and there are two possible responses. Either the teacher presents the scientific perspective and ignores all other views, or the teacher provides opportunities for both scientific and everyday views to be considered.

Returning to a view of learning that involves being introduced to a new social language, we would argue that a fundamental part of this process is to see how different perspectives overlap and articulate with one another. From a science teaching point of view, this involves explicitly dealing with those issues that seem to sit uneasily between everyday and scientific views. In this way the 'misconceptions' identified in the children's thinking literature become sources of key questions to be addressed in teaching: 'What lies between the air particles?'; 'Do the particles expand when the solid is heated?'; 'What do I mean when I talk about a model for gases?' Of course, such an approach is intellectually (and emotionally!) demanding for both teacher and students. In these lessons, we can very easily imagine Pedro's heart sinking as the students seem to *accept* that there is nothing between the particles, but *then* use this as evidence for particles expanding. We also think back to Raquel and Alex being driven 'crazy' in talking through a general relationship between particle energy and motion. Talking and thinking in this way is often very demanding. Nevertheless, we would strongly maintain that it is precisely at these moments, when the dialogicality of the meaning making process (with views being placed alongside each other) is most evident, that there is the greatest chance of significant learning occurring.

### Students 'talking science' and 'playing teacher'

A further feature of this teaching and learning performance is the amount of time allowed for the students to work together in small groups. Throughout successive episodes, one cannot help but be impressed by the high quality of the discussion (certainly in the group we observed), and the way in which the students stuck to their task. By the end of the lessons, the students were able to talk the school science *social language* fluently, and we take this as direct evidence of personal sense making and learning having taken place. In addition, there is evidence of the students becoming aware of various features of the nature of scientific knowledge (or the grammar of science) as they developed explanations that, for example, must fit in with other ideas and be logically consistent: 'it has to be logical!'

Alongside the students' coming to terms with the school science view, we have seen them becoming fluent in the *speech genre* of school science. Thus, Carolina took on the role of teacher very effectively, both in keeping her fellow group members on task and in reviewing the group's thinking, by means of an I–R–E confirmatory exchange. Such interventions are an important help in maintaining progress in small-group situations without the direct involvement of the teacher. Later, another student, Fabiola, intervened in a whole-class situation to evaluate a fellow student's point of view and to emphasize that 'the particles themselves don't expand' on heating, this time enabling the whole class to progress with the scientific story.

In Chapter 4 we were able to identify a clear pattern in the classes of communicative approach used by the teacher in staging the teaching and learning performance. From the analysis of these lessons, it is clear that such variation applies just as well to groups of students working together. At times the group talk was authoritative in nature, as the students focused on one perspective; at other times it was heavily dialogic, as the students worked together to generate new meanings. It is clear that the discourse of science classrooms can be authoritative not only when the teacher is taking the leading role, but also when the students are talking among themselves.

The general point here relates to the importance of providing opportunities for students to *talk their way* into the science story. This applies to all the major phases of the teaching sequence, whether developing the scientific story, assisting the students in coming to understand the story or supporting students in applying that scientific point of view. By allowing these opportunities to talk through the science and to relate it to everyday views, we believe that there is a greater chance of students mastering the scientific perspective and making it their own. Perhaps there will also be a greater chance of people coming to understand and to appreciate the latest advances in science and technology, even beyond the possibilities of travel to the Moon!

# 6 Looking back . . . looking forward

A science teacher friend recently completed a sequence of lessons on the particulate theory of matter. He had spent quite a lot of time exploring students' points of view, addressing particular key barriers to learning and allowing enough time for the students to talk and work through the ideas for themselves. In fact, he took an approach that was very similar to that outlined in Chapter 5. In the very next lesson with this class of 14-year-olds he was due to teach the concept of 'speed'. For this lesson, our friend briefly reviewed the idea that speeds are calculated as distance per unit time, and then set the class 20 questions to work through in silence. At the end of the lesson one of the students, a boy, came up and commented, 'That was great sir, just getting on with it, a nice change from all that talking!'

In this final chapter, we move from the specific cases of the two teaching sequences to consider the more general messages for teaching and learning science that emerge from this book. First, we return to thinking about the five aspects of the analytical framework. We then look forward to consider how those aspects might be used as a basis for planning teaching and further to consider how they might be introduced to teachers in professional development contexts. Finally, we focus on the students to reflect on the nature of *their* experiences while they were actively involved in these teaching and learning performances.

## The analytical framework

In Chapter 3 we introduced the five aspects that constitute our analytical framework. At the heart of the framework lies the *communicative approach*, which characterizes how the teacher and students interact in working on ideas and understandings. Four classes of communicative approach are identified,

ASPECT OF ANALYSIS

| FOCUS | 1 Teaching purposes | 2 Content |
|---|---|---|
| APPROACH | 3 Communicative approach | |
| ACTION | 4 Patterns of discourse | 5 Teacher interventions |

**Figure 6.1**    The analytical framework: a tool for analysing and planning science teaching interactions.

based on the *interactive–non-interactive* and *dialogic–authoritative* dimensions, with the former focusing on the degree of interaction between classroom participants and the latter on the diversity of points of view that are taken into account in the classroom discourse.

The *teaching purposes* and *content of the discourse* relate to the *focus* of the classroom talk. Six teaching purposes, based on sociocultural theory and developed through classroom observations, are identified. The analysis of content is made in terms of the distinction between everyday and scientific social languages, and the categories description–explanation–generalization (both empirical and theoretical).

The *patterns of discourse* and *teacher interventions* focus on the *action* of the science classroom, in providing the tools for directly analysing the patterns and forms of the classroom talk. The patterns of discourse are the distinctive and regular patterns of interaction that emerge as teacher and students take turns in classroom talk. The teacher interventions are the ways in which the teacher works to develop the scientific story and to make it available to all of the students.

Having applied these five aspects of the framework to analysing the two teaching sequences, we now return to discussing them more broadly. First of all, we consider the impact of the *content* on the development of a sequence of science lessons, and then try to identify any patterns in the ways in which the other aspects interrelate, one with another.

### The impact of content

In many linguistic studies of classroom discourse the analyses are carried out, and the findings reported, solely in terms of patterns of interaction, and the actual *content* of what is being taught and learned is not regarded as being a significant feature. Thus the content of the lessons is often taken as being simply a contextual factor (whether on 'volcanoes' or 'democracy' or 'particle theory') that is not addressed in the analyses of the interactions and the way in which the lessons develop.

We see things differently. Looking back over the analyses of the two lesson sequences, presented in Chapters 4 and 5, we consider that perhaps the most striking difference in the ways in which the teaching performance developed in each case followed from the differences in the nature of the *content* addressed in each sequence. Thus, in the rusting case, the aim of the lessons was to end up with a *generalized empirical description* of the phenomenon of rusting, namely that water, air and iron are the substances essential for this chemical reaction. On the other hand, for the particles lessons, the aim was to develop a *theoretical* model, to be *generally* applied in *explaining* a full range of physical properties of matter. As discussed in detail earlier, the differences in the nature of the content to be addressed led to a relatively smooth passage from everyday to scientific views with the rusting lessons, but to a passage that was consequent upon significant learning barriers for the particles case.

Although both teachers planned their teaching to explore and take account of the students' views, the differences in the relationship between everyday and scientific perspectives had an obvious impact on the development of the teaching and learning. Using the terminology introduced in Chapter 2, we would argue that a significant difference in *learning demand* (Leach and Scott 2002) contributed to the difference in the ways in which the teaching performance evolved in each case.

Thus, we see the scientific content to be addressed, and more specifically its relationship to everyday views, as being a key factor to be taken into account in analysing or planning any teaching and learning sequence. In a strong sense, the scientific content sets the scene for all that follows. Furthermore, we are convinced that the respective distinctions between everyday and scientific social languages and description–explanation–generalization offer a powerful basis for thinking and talking about the impact of content on teaching and learning science.

### How the different aspects fit together

What can we say, in *general* terms, about the ways in which the different aspects of the framework fit together as a teaching performance develops? Is it the case, for example, that specific teaching purposes are associated, theoretically and practically, with particular communicative approaches? Are we in a position to say anything about the ways in which different communicative approaches might be developed through different patterns of discourse? These are the sorts of questions that we address in the following sections and that we regard as being of central importance in further demonstrating the coherence of the framework.

### From purposes to approach

What can we say about how different *teaching purposes* might be addressed through any of the four classes of *communicative approach*? In both the rusting and particles lesson sequences the teacher started with an activity to explore students' views and, following on from this activity, we observed each teacher engaging the students in cycles of *exploring students' views, working on students' views, maintaining the scientific story*.

In exploring the students' views, both teachers adopted an *interactive/ dialogic* communicative approach, in which they helped the students to clarify their own ideas, accepted them and made them available on the social plane. In working on those ideas, the teachers developed an *interactive/ authoritative* communicative approach, in which the teacher selected some contributions and discounted others, marking key ideas and checking student understandings. In maintaining the scientific story, the teachers reviewed and summarized the key points through a *non-interactive/authoritative* communicative approach.

The combinations of teaching purpose and communicative approach that we see here make sense to us. If the teacher wants to explore students' views they need to adopt an interactive/dialogic approach; if they want to work on students' views then an interactive/authoritative approach suggests itself. We are not, however, arguing that there should *always* be such a direct relationship between purpose and approach. Teaching never works out in that precise, predictable kind of way in practice. For example, it would be perfectly possible to act to maintain the scientific story through an interactive/ authoritative approach, with the teacher rehearsing the progress achieved through interactions with students, rather than by making an authoritative presentation.

The point that the same teaching purpose might be effectively addressed through different communicative approaches is borne out by the way in which the scientific point of view was introduced in each of the two teaching sequences. In the rusting case the teacher was able to follow the largely *inter- active* teaching cycle set out above, including both *dialogic* and *authoritative* phases, in moving from the students' views to the school science point

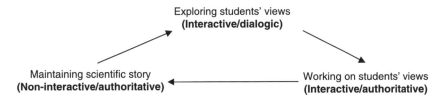

Exploring students' views
**(Interactive/dialogic)**

Maintaining scientific story
**(Non-interactive/authoritative)**

Working on students' views
**(Interactive/authoritative)**

**Figure 6.2**   A teaching cycle.

of view. In the case of the particles lessons, with the quite different relationship between everyday and scientific views, the teacher initially explored the students' points of view, through an interactive/dialogic approach, and then introduced the scientific perspective through a *non-interactive/ authoritative* presentation. Even though there were these differences in approach to introducing the scientific view, it is important to recognize that *both* involved authoritative interventions from the teacher, and indeed this is only to be expected. Introducing students to the scientific social language will inevitably involve some form of authoritative intervention by the teacher.

For the final two teaching purposes, *supporting internalization* and *guiding students to apply the scientific view*, the communicative approach taken is likely to be almost exclusively *interactive* in form. Here, the teacher moves back and forth along the dialogic–authoritative dimension, both probing the students' current understandings and providing specific points of information and guidance.

### From approach to action

It is one matter to consider how specific teaching purposes might be addressed through particular communicative approaches, and quite a different kind of problem to consider how a particular communicative approach might be enacted in the classroom. For example, we might agree that it is appropriate to address the teaching purpose of exploring students' views by means of an interactive/dialogic approach. The key question then becomes one of *how* the teacher actually develops such an approach with their students. With regard to this issue, we believe that there is a very important relationship to be explored between *patterns of classroom discourse* on the one hand and *classes of communicative approach* on the other.

In our analyses of the two lesson sequences, we identified two main patterns of discourse (Table 6.1). How do these two patterns of discourse relate to specific communicative approaches? First, there is a clear link between the *I–R–E* pattern of interaction and an *interactive/authoritative* communicative approach. This relationship is illustrated by the episode presented in Chapter 3, 'Let's just ignore the sparks', where the teacher moves towards a particular point (that the bell heats up) through a pattern of interactive triads. Furthermore, in the rusting case Lynne established that 'cold' is not essential for rusting through an I–R–E pattern of interaction, while Pedro used the same triadic pattern to make the point that there is 'empty space' between the particles.

Second, there is a link between the *I–R–F–R–F–* chains of interaction and an *interactive/dialogic* communicative approach. Here the teacher explores different points of view by engaging the students in extended sequences of turn-taking. Thus we saw Lynne, in the rusting case, engaging Rebecca

**Table 6.1**  Two patterns of discourse

| Pattern of discourse | Form |
| --- | --- |
| Initiation–response–evaluation (I–R–E) pattern | Triadic |
| Initiation–response–feedback (I–R–F–R–F–) repeated pattern | Chain |

in a chain of interaction as she tried to find out more about Rebecca's experiment. In the particles case we saw similar chains of interaction, this time as the *students* talked through the relationship between energy and particle motion.

These two relationships between communicative approach and patterns of discourse are summarized in Figure 6.3. These links between pattern and approach follow logically from our definitions of the aspects and are borne out by examples from the classroom data. At the same time, we must re-emphasize the point that these can never be unique and exclusive relationships. Thus, while I–R–E triads are linked to an interactive/authoritative approach, and chains of interactions tend to lead to an interactive/dialogic communicative approach, we have found exceptions to these rules. For example, in the first episode of the rusting lessons the teacher established an *interactive/dialogic* communicative approach, exploring and taking account of students' ideas, at least in part through a series of *triadic* interactions. In this particular case, however, the triads were not of the classic I–R–E form. As the teacher collected ideas from the students, she did not evaluate their ideas, taking into account the scientific view on these matters, but simply accepted and collected them on the board. In this way the discourse followed a triadic pattern, but did not involve the sharp evaluative steps that are associated with the interactive/authoritative approach.

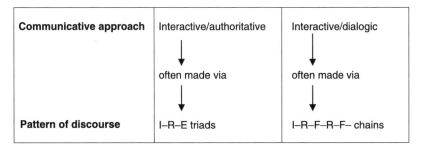

**Figure 6.3**  Relationship between communicative approach and pattern of discourse.

*Fitting it all together*

Having identified some clear links between approach and purposes, and between approach and patterns of discourse, we begin to get a sense of how the whole framework fits together. Specific teaching purposes can be addressed through particular communicative approaches, and these approaches are put into action through different patterns of interaction and teaching interventions.

For example, we might take the teaching purpose of *working on students' views*. As outlined above, this purpose might be addressed through an *interactive/authoritative* approach, based on *I–R–E triads*. Furthermore, in working on the students' views it is quite likely that the teacher will make interventions to *select ideas* or *mark key ideas*, possibly by means of a *confirmatory exchange*.

So far in this book we have focused our attention on developing a framework to be used in *analysing* science teaching. In exploring the various links between aspects, however, the potential for using the framework as a basis for planning teaching becomes increasingly evident. We now turn, therefore, to considering the other side of the coin and to addressing the question of what we can say about planning and implementing science teaching.

## Planning and implementing science teaching

A central feature of the view of teaching and learning science that we outlined in Chapter 2 is that meaning making always involves bringing together, and working on, different ideas, and is therefore *dialogic* in nature. It is through students comparing and contrasting their views with the scientific one that they can begin to make sense of the scientific story being taught. Arguing more broadly, if school science is to prepare students to be citizens who are critically aware of the different perspectives that are at issue when a societal problem has a scientific component, then the science taught in school must be related to common-sense, everyday views. The only way that this can be achieved is through dialogue, which results in students enlarging their already heterogeneous cultural views, with science offering one more perspective to be added to the 'toolkit' that students can draw upon. If one accepts these views of what is involved in learning science (or anything else for that matter), then it is clear that teaching sequences must provide opportunities for teacher and students to work on ideas with a dialogic approach.

At the same time, and as argued earlier, if students are to learn the social language of science, then somewhere within the teaching and learning performance there must be an authoritative introduction to the scientific point of view. Students will not stumble upon, or discover, the key concepts of the social language of science for themselves. It follows, therefore, that there will always be a tension between dialogic and authoritative discourse, and a key part of the science teacher's role is to strike an effective balance between dialogic and authoritative communicative approaches.

The general message is that we would expect any effective teaching sequence to include passages of *both* dialogic and authoritative discourse, achieved in *both* interactive and non-interactive ways. In other words, we are committed to the idea that any teaching sequence should include activity within all four classes of communicative approach, and that as the teaching performance unfolds there should be movement between the four approaches.

### Planning the way in which the teacher interacts with students

The idea that teaching sequences should be *planned* to engage teacher and students in different classes of communicative approach is *not* one that is normally encountered in current science teaching practices. As we pointed out in Chapter 1, present-day approaches to planning science teaching tend to focus on developing sequences of *activities* for the students, and no explicit attention is given to the nature of the talk around the activities, the talk through which the scientific story is developed. We believe that this is a significant shortcoming.

What is needed is an approach to planning in which explicit attention is paid to the nature of the talk in each phase of the lesson sequence and how that links to both the teaching activities and the teaching purposes. For example, in addressing the purpose of *guiding students to work with scientific ideas*, thought needs to be given not only to what the students will be doing (the teaching activities), but also to the ways in which they can be encouraged to talk through the scientific view for themselves. In the particles teaching sequence, we saw many impressive examples of how this was achieved by organizing the students to work in small groups on clearly framed tasks.

In this way we see lesson planning as fundamentally involving the integration of *teaching purposes*, *teaching activities* (teacher demonstrations, student practical activities, small group discussion activities and so on) and *communicative approaches*.

Lesson planning thus involves identifying both teaching activities and communicative approaches to address specific teaching purposes. Furthermore, just as teachers are familiar with including a range of different kinds of teaching activities in a lesson sequence, so too there is a need to vary the

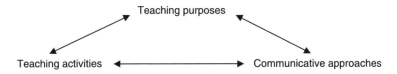

**Figure 6.4**  The integration of purposes, activities and approaches in lesson planning.

classes of communicative approach. If, for example, the teacher introduces the scientific point of view through an authoritative presentation, an important question in planning the *next* phase of the sequence is how the students can now be given the opportunity to talk through, and to explore, this point of view for themselves. In this way the teacher establishes an appropriate 'rhythm' to the discourse.

### Taking account of the content in planning

Returning to the point made earlier, a further key aspect to be considered in the planning process is the content of the teaching and learning performance, and in particular the relationship between the school science and everyday points of view. What does this mean in practice?

Put simply, if the *learning demand* is considered to be high for a specific topic area, then the lessons need to be planned to allow plenty of time for teacher and students to talk through these new and unfamiliar ideas. On the other hand, if the learning demand is deemed to be slight, less time will need to be spent on the topic, as the students are more likely to find that 'it's obvious' or that 'it makes sense'. In this way the very different approaches taken by our friend (referred to at the start of the chapter) to teaching 'particles' and 'speed' make a lot of sense, and were also clearly approved of by at least one of his students!

### An approach to planning science teaching sequences

Elsewhere (Leach and Scott 2002) an approach to planning science teaching, based on these ideas, has been developed. The approach involves four linked steps:

1  Identify the area of the *school science social language* to be taught.
2  Consider how this area of science is conceptualized in the *everyday social language* of students.
3  Identify the *learning demand* by appraising the nature of any differences between 1 and 2.
4  Develop a teaching sequence to address each aspect of this learning demand, identifying: (a) the *teaching purposes* for each phase of the sequence; (b) how those purposes might be addressed through an appropriate *communicative approach*; (c) how this approach might be put into action through appropriate *teaching activities*, *patterns of discourse* and *teacher interventions*.

The first three steps address the impact of content and the fourth moves from purposes to approach to action.

## But does this lead to more effective teaching?

We are developing a line of argument here that calls for quite a fundamental reappraisal of how we conceptualize the planning and implementation of science teaching. What evidence do we have to support the view that such an approach might be more effective than existing practice in supporting student learning?

A small number of studies do exist in which researchers, drawing on ideas from the sociocultural tradition, have demonstrated that diversity in patterns of discourse, and the presence of dialogue in particular, might be associated with enhanced learning outcomes. For example, Leach *et al.* (2003) have carried out research in this area as part of the ESRC sponsored 'Evidence Based Practices in Science Education' project. In this research project, the team based at the University of Leeds worked with small groups of local science teachers to develop three short teaching interventions, which were planned following the four-step approach summarized in the previous section. Thus, explicit attention was paid to identifying the learning demands generated for each topic (simple electric circuits, plant nutrition, particle theory). Lesson sequences were then developed to address those demands, clearly identifying the teaching purposes for each phase and addressing those through appropriate combinations of activity and approach.

The teaching sequences were implemented, by the teachers, as part of the normal science curriculum in their own schools. The effectiveness of the sequences was investigated by making measures of student learning gains (established through pre- and post-teaching diagnostic tests), which were then compared with the gains achieved with parallel classes following the school's normal curriculum. It was found that for all three schemes the overall learning gains (particularly in relation to questions probing understanding) were greater for those classes following the 'new' schemes than for the comparison classes. The authors acknowledge that the research did not follow a strict experimental design, with careful control of 'variables' between experimental and comparison groups. This was not practicable since the work was implemented as part of each school science department's ongoing teaching. Nevertheless, the trend in comparative gains was seen in the implementation of the schemes across nine different high schools. Of course, the issue that this kind of study cannot address is that of identifying which aspects, or combinations of aspects, of the new teaching interventions (The identification of teaching purposes? The teaching activities? The focus on communicative approach?) were influential in generating the enhanced learning gains.

The two teaching sequences presented in this book also include the kinds of approaches that we are arguing for here, and this, of course, is no coincidence, since these particular sequences were chosen for this very reason.

We have evidence to suggest that both sequences were effective in supporting student learning. For example, after the particles lessons, the students were tested, and 70 per cent of them were found to be able to use all the features of the particles model introduced in the lessons in providing explanations of both familiar and new properties of matter.

Perhaps just as compelling an indicator of student learning in the particles lessons is the evidence provided by the group work transcripts, which demonstrate the impressively sophisticated ways in which the students were able to engage in talking about a range of aspects of particle theory and its application. This is true for both the earlier episodes of the sequence in which the students were struggling to come to terms with specific aspects of the scientific model, and the later phases in which they talked through explanations confidently and precisely, applying the scientific point of view. It is clear that the students were only able to engage in these sustained learning dialogues because the teacher directed the teaching performance in this way, leaving space and time for such activity.

## From theory to practice: teacher professional development

In the first chapter of this book, we made the comment that lots of the science lessons that we see in schools are limited in terms of the kinds of teacher–student interactions, and that by far the most common is the interactive/authoritative approach. We also suggested that lots of science teachers adopt this form of presentational style simply because it represents the existing, invisible, taken-for-granted practice of science teaching. So how might we encourage and support science teachers to extend their practice, following the kinds of approaches set out above?

Perhaps the first point to be made is that reviewing and developing professional practice in this way can be challenging to the individual. As in learning science, the level of challenge depends on the size and nature of the gap between the teacher's existing practices and what is being asked of them. Put simply, if the teacher feels most confident when taking an authoritative approach, in which the scientific view is presented as factually 'given', then they will certainly feel challenged by an approach that involves open discussion of the students' views and understandings. At the same time, there will be other teachers for whom the notion of cycling around the classes of communicative approach seems obvious because it is very close to what they already do. Either way, what might be done to support teachers in extending their teaching practice?

Part of the job here is to make links to the fundamental ideas of what is involved in teaching and learning science (as discussed in Chapter 2) in order that a convincing rationale can be developed for the practical teaching approaches being suggested. In addition, it is clear that a set of tools, or vocabulary, is needed for talking and thinking in a meaningful way about these aspects of teaching practice. We have found the analytical framework to be extremely helpful, in this respect, while working with science teachers in both pre-service and in-service professional development contexts.

For example, in the context of pre-service teacher development courses in Brazil, we have used the framework as a basis for analysing videotapes of the student teachers teaching science in one of their practice schools. In analysing the videotapes during university sessions, the student teachers are first asked to work in small groups to make a map of the main activities of the lessons, and to divide these up into the main teaching episodes that constitute each lesson. The students then decide what is the purpose of each episode, using the list of six teaching purposes presented in the framework. Having identified the different episodes and associated purposes, the students analyse each episode in terms of the different aspects of the framework. The following questions are set to guide their work:

1   *Content*. What kind of content (everyday or scientific views; empirical or theoretical; descriptions, explanations or generalizations) is most prominent in the episode? Is there any movement in the nature of the content during the episode? (Perhaps from: everyday views to the introduction of a scientific model; describing to explaining a particular phenomenon).

2   *Communicative approach*. What kind(s) of communicative approach is used by the teacher in each episode? Is one specific class of communicative approach particularly prominent?

3   *Patterns of discourse*. What are the patterns of interaction that develop during each episode, as teacher and students take turns in the classroom talk?

4   *Teacher interventions*. How does the teacher intervene, during each episode, to develop the scientific story, and to make it available to all the students?

Having carried out this analysis, each group is asked to choose two or three episodes that they consider to represent examples of good and bad teaching practice. Video clips of these episodes are then presented, along with their analysis, to the whole class. In the subsequent discussion, alternative teaching approaches are considered and the ways in which those approaches might be played out in practice in the classroom are talked through in detail. Our experience suggests that the student teachers can learn a great deal from this kind of discussion. This is especially so in helping students to progress from recognizing the need for a certain approach ('it would be good to have the students talk through their views at this point') to having an idea of how this might be achieved in practice ('what I would do is . . . ').

For example, one striking feature of the teaching in the rusting and particles cases was each teacher's ability to *sustain* chains of dialogue with students when developing interactive/dialogic sequences. Thus, both teachers were able to encourage students to keep talking, providing them with the opportunity to elaborate upon, and to clarify, their points of view. They often achieved this simply by repeating parts of what the students had just said, or offering low-key comments of encouragement, as the element of feedback in the developing I–R–F–R–F– chain. This ability to support students in talking on-task, and to take account of the students' points of views, is not to be underestimated. At times both teachers were so effective in prompting and sustaining student participation that they were able to withdraw altogether for a number of turns, as students responded to, and provided feedback on, the comments of other students. Indeed, on some occasions the students assumed control to the extent that they took the lead in setting the developing agenda by raising new questions and issues for discussion. Our experience suggests that this ability to sustain chains of dialogue is something that student teachers can learn to do, once it has been drawn to their attention and the underlying teaching purposes have been discussed. Just as teachers can learn how to use different teaching activities, so too can they learn how to develop different kinds of teaching interaction in the classroom.

Following these initial presentations and discussions, the students carry out further teaching in their practice school and later in the year the whole cycle of activity is repeated, with the analysis of the second tape leading to an evaluation of any *changes* in teaching practice for each student.

At the end of the course the student teachers complete an evaluation questionnaire for the programme. Because of the shortage of chemistry teachers in Brazil, some of the students already have experience of teaching chemistry prior to graduation. In the evaluation, these particular students are asked if they have changed any aspects of their teaching as a result of what they have seen and done during the course. The majority of these students comment that they have become more aware of their own teaching, particularly with regard to the different teaching purposes and how they can use different communicative approaches to address those purposes. In addition, they often comment that they have changed the way in which they interact with their students in school, valuing the students' ideas and contributions more, and becoming more skilled in checking the understanding of their students. The following written comment from one of the students illustrates this conscious awareness of the teaching process that the analytical framework has helped to promote:

I began teaching chemistry one year ago. I found it interesting to recognize that some of the video situations we discussed during the course repeated themselves in my classes in school, and that these situations could have different outcomes. The work we did during the course has helped me to be more critical about these situations and, more importantly, to think about these situations and to plan how to cope with them more effectively. It also helped me to think about what to do when introducing different concepts to the class. I have learned to hear what the students have to say from their own points of view. My classes are now much more interactive and alive and I have a better sense of the learning outcomes as the students are more confident in asking questions and expressing their understanding. I have learned how to make interventions to make their ideas available to the whole class, how to check their understandings, how to review the progress made in the lessons.

## Back to the students

In this final section, we turn from these teaching matters to reflect more broadly on the experiences of the *students* as they took part in the two lesson sequences.

### Developing an understanding of the scientific view

As argued above, there is evidence from both of the teaching and learning cases to suggest that the students made good progress in coming to understand, and to be able to apply, the school science view. What we would like to consider now, in more general terms, is the way in which the students moved from an initial position of knowing very little about the scientific subject matter, to a final state of understanding it quite well. To do so, we return to some ideas proposed by Bakhtin.

Bakhtin (1934) suggests that the gradual appropriation of meanings by individuals follows a path that starts with the 'new ideas' being introduced to the social plane. This initial stage of appropriation maps on to the early phases of the teaching sequence, where the teaching purposes of *opening up problems*, *exploring students' views* and *introducing and developing the scientific story* are addressed. During this stage the ideas are very much seen by the individual as being foreign to their own experience and belonging to 'others'.

The next phase in the progressive appropriation of meaning, according to Bakhtin, is reached when the individual begins to see the new idea as half their own and half belonging to 'others'. This point is likely to be achieved as the teacher addresses the teaching purpose of *guiding students to work with scientific ideas and supporting internalization*, or, in other words, as the teacher helps the students to make individual sense of the school science ideas. As each student begins to use a school science idea that is new to them, they often do so in a faltering and uncertain manner, which indicates that the idea has not yet become fully internalized. The new way of thinking and talking is still only half their own.

The final phase described by Bakhtin along this path of appropriation of meanings occurs when the idea is fully appropriated by the individual. This is likely to occur as the teacher *guides students to apply the scientific view and hands over responsibility for its use*. Thus, as the student applies the scientific view to a range of different phenomena and situations, they are able to make the scientific meanings their own, thereby appropriating the scientific view. We saw evidence, particularly from the particles sequence, of the students using the school science ideas in quite a sophisticated way to develop their own arguments. For example, near to the end of the sequence Alex read aloud his written response to a question on the states of matter:

> The particles in the solid and liquid states are grouped together and because of that they have a certain consistency, and their volume does not change. Since the average distance between the particles of a gas is about 10 times bigger in magnitude than the separation of the particles in a liquid, we know that the particles of the gas are more spread out and can move around freely, and thus the gas has no definite volume.

It is through working with, and applying, the scientific ideas that the students can build upon and elaborate their individual understandings, to achieve this kind of fluency in the scientific social language.

So, the sequence of stages of appropriation of meanings proposed by Bakhtin captures the gradual nature of the students' learning, which was particularly apparent in the analysis of the particles lessons. Furthermore, the teaching purposes identified in the framework fit easily with this picture of progressive appropriation (Figure 6.5).

### Learning science, learning the speech genre of school science

One of the striking findings from the analyses of the small-group work during the particles lessons was the ways in which the students, themselves, generated the triadic and chained patterns of discourse that we would normally expect to originate with the interventions of the teacher. Thus, some students assumed the role of teacher when working in small groups, as they asked for contributions from fellow students, gave feedback on those contributions and then moved on to raise further questions. As we saw in the

**Stage 1 of appropriation**

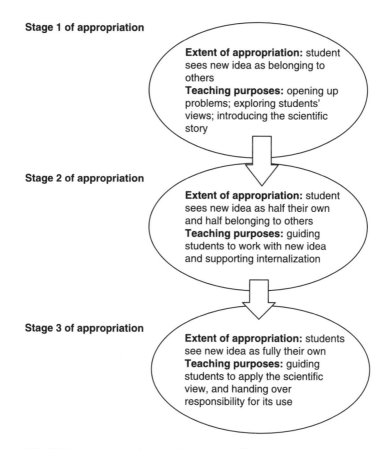

**Extent of appropriation:** student sees new idea as belonging to others
**Teaching purposes:** opening up problems; exploring students' views; introducing the scientific story

**Stage 2 of appropriation**

**Extent of appropriation:** student sees new idea as half their own and half belonging to others
**Teaching purposes:** guiding students to work with new idea and supporting internalization

**Stage 3 of appropriation**

**Extent of appropriation:** students see new idea as fully their own
**Teaching purposes:** guiding students to apply the scientific view, and handing over responsibility for its use

**Figure 6.5**  Linking the stages of appropriation to teaching purposes.

particles lessons, these authoritative teaching interventions *by students* were very effective in supporting the development of group discussions, keeping the talk focused on the task at hand. At other times, the talk was more dialogic in nature as the students engaged in a genuine exploration of ideas, offering and working on different points of view, as they tried to make progress with questions such as 'how is energy related to the particle motion?'

It is clear that, in their science lessons, the students were learning more than just the content of school science; they were also learning the speech genre of school science and drawing on it to make progress with their small-group work. The obvious point to be made here is that if students are to draw upon these different communicative approaches in school science, then those ways of talking need to be *modelled* by the teacher in work with the whole class. If, for example, the teacher consistently takes an interactive/ authoritative communicative approach to teaching science, then it is likely to be difficult for students to engage in interactive dialogue about science questions, simply because they have little or no experience of such ways of talking about science. The likely outcome from such a situation is that the group work will seem like a 'waste of time' to the students, who just want the teacher to tell them the 'right answer'.

Even more fundamentally, if the talk of the social plane is restricted to the authoritative classes of communicative approach, then those ways of *talking* about science are likely to be internalized as the principle ways of *thinking* about science. If we want to encourage students to engage in explorative thinking about scientific matters, taking into account different points of view and trying to make links between them, then this mode of *thinking* needs to be modelled in the *talk* of the social plane.

### The emotional response of the students

So far we have considered how the students responded to the two lesson sequences, in terms of what they learned, the gradual and progressive nature of that learning and the features of their talk in small groups. What can we say about the emotional response of the students to the lessons?

In both lesson sequences, the students responded enthusiastically to the teaching approaches taken. Going back to the ideas introduced in Chapter 2 (Arruda *et al.* 2001), it is clear that the students engaged *actively* in the lessons, taking part in classroom debates, asking questions and becoming involved in the activities prescribed by the teacher. Indeed, in some situations they took the *initiative* in developing ideas beyond what was required by the teacher. This active engagement was apparent in the way in which the students contributed to discussions in both whole-class and small-group contexts, often being prepared to offer their points of view spontaneously. It was also apparent in the way in which the students stuck to the task of talking through difficult problems, overcoming their frustrations to project a real sense of achievement on successfully reaching a solution.

Why did the students respond in this positive way? We believe that they did so, at least in part, because the teachers in both cases contrived to develop a teaching and learning context in which they consistently probed, and took into account, the students' ways of thinking. In each case, the 'scene was set' by the teachers opening up the problem and actively exploring the students' initial ways of thinking. This was not a token exercise. The students' ideas were recorded and taken into account in the subsequent development of the scientific story. Right from the start of the lessons, the students therefore recognized that their voices would be heard, and that there was room for dialogue between their points of view and the scientific view. In this atmosphere the students were quite happy both to offer their opinions and to engage in making sense of the scientific view. We believe that the teachers got the balance between authoritative and dialogic approaches just right, in being able to draw the students and their ideas into the discussions, while also providing the authoritative interventions needed to maintain the development of the scientific story.

## Final words

We have observed this kind of positive intellectual and emotional response from students on many occasions when they have been exposed to the kinds of teaching approaches we have outlined in this final chapter. During one such sequence of lessons on the particulate theory of matter (see Johnston and Driver 1991) a student wrote in her informal 'learning diary':

> The lesson was quite hard, because we had to find out what was in the middle of a solid, liquid or gas. I thought about particles. If you say that the particles are spread out like this in a gas, then what would be between them? I thought that this is a very hard question.

Another student wrote:

> Overall I think I enjoyed this new style of teaching. I suppose the course made me more confident to put forward my ideas . . . We were able to think more for ourselves and we were able to put forward our ideas, and each idea was discussed. If I had a choice of which lessons to do, it would be the recent ones.

Another wrote:

> On the whole it was a very interesting lesson. It's not very often in class discussion when somebody says something and nobody laughs at what that person has said.

Reflective intellectual engagement, enhanced self-confidence and com-mitted emotional involvement, in teaching and learning about the *particulate theory of matter*! What more could a teacher want? When the third student, Hazel, wrote in her diary that 'it's not very often when . . . nobody laughs at what that person has said', she was referring to the fact that no one laughed at *her* in those science lessons, something that occurred frequently in other classes. We have no doubt that all these outcomes followed from the way in which the teacher conducted those lessons, listening to the students' points of views, probing their understandings and encouraging each student to do likewise. In those science lessons the students not only learned something about the particulate theory of matter and the speech genre of school science, they also began to develop a basic respect for and interest in the views of others.

It is with these kinds of thoughts in mind that we point to the value of the approaches to planning and implementing science teaching that we have set out in this final chapter. It is with these thoughts in mind that we recommend to you the analytical framework that we have introduced in this book as providing a set of fundamentally important tools for thinking and talking about science teaching and learning.

# Appendix: Further notes

In this appendix we offer further discussion of the key theoretical and methodological issues that frame the analysis presented in the book. This additional material is numbered in the text, and is aimed at those readers who wish to probe a little more deeply into some of the key ideas.

## Note 1: Vygotsky's general genetic law of cultural development

Vygotsky formulated his ideas about the passage from social context to individual understanding in terms of his 'general genetic law of cultural development':

> Any function in the child's cultural development appears twice, or on two planes. First it appears on the social plane, and then on the psychological plane. First it appears between people as an interpsychological category, and then within the child as an intra-psychological category. This is equally true with regard to voluntary attention, logical memory, the formation of concepts, and the development of volition. We may consider this position as a law in the full sense of the word, but it goes without saying that internalisation transforms the process itself and changes its structure and functions. Social relations or relations among people genetically underlie all higher functions and their relationships.
>
> (Vygotsky 1931: 163)

In this way, Vygotsky outlines the social origins of 'higher mental functions'. It is important to recognize, however, that according to Vygotsky these mental functions, such as memory, attention, perception and thinking, *first* appear in an *elementary* form, as the result of a *natural* line of development.

> We shall call the first structure *elementary*; they are psychological wholes, conditioned chiefly by biological determinants. The latter structures that emerge in the process of cultural development, are called *higher structures*. The initial stage is followed by that first structure's destruction, reconstruction, and transition to structures of higher type. Unlike the direct, reactive processes, these latter structures are constructed on the basis of the use of signs and tools.
>
> (Vygotsky 1978: 124)

It is only through social interaction and participation in cultural life that these elementary functions are transformed to higher mental functions. While the former can be found in both humans and other animals, the latter are unique to humans. According to Wertsch (1985: 25), Vygotsky used four criteria to distinguish between elementary and higher mental functions:

1   The shift of control from environment to the individual; that is, the emergence of voluntary regulation.
2   The emergence of conscious realization of mental processes.
3   The social origins and social nature of higher mental functions.
4   The use of signs to mediate higher mental functions.

An important feature of Vygotsky's perspective is that, although social processes are not simply transferred from the interpsychological to the intrapsychological plane, there is an inherent connection between the two planes, as the means that mediate higher mental functions, such as language, are essentially social in nature. Wertsch (1991: 14) refers to this feature when he states that the 'mind extends beyond the skin'.

## Note 2: Sociocultural perspective

In using the term *sociocultural* to characterize the theoretical perspective that informs this book we are following the lead of James Wertsch (1991), who suggests this term instead of *sociohistorical*, which was used by Vygotsky himself and his colleagues. The basic assumption is that all mental actions (such as learning science) are inevitably situated in cultural, historical and institutional settings and what is accepted as knowledge and teaching and learning in school science is clearly related to these settings. What constitutes the shared knowledge and activities of a community of science teachers or scientists, in one period of history, is influenced by the historical, cultural, social and economic conditions of that period.

For example, it is no coincidence that in the first decade of the twentieth century all aspects of cultural life – music, performing arts, science, politics and so on – found themselves in the middle of a revolution that would completely change Western culture and the way that ordinary people led their lives. We can understand these changes only if we examine the dynamics of the historical, cultural and social processes that contributed to their emergence. Similarly, we can only come to understand the processes of teaching and learning in a specific school if we are able to examine the social, cultural and historical contexts that frame them.

## Note 3: Bakhtin's views on dialogue and dialogism

Dialogue is, according to Holquist (1990: 15), 'an obvious master key to the assumptions that guided Bakhtin's work throughout his whole career'. Dialogism implies recognition of the importance of *the other* both for existence and for language. Existence is dialogic in the sense that any individual consciousness is formed against the background of others, and of social practices such as language:

> The world addresses us and we are alive and human to the degree that we are answerable, i.e., to the degree that we can respond to addressivity. We are responsible in the sense that we are *compelled* to respond, we cannot choose but give the world an answer. Each one of us occupies a place in existence that is uniquely ours; but far from being a privilege, far from having what Bakhtin calls an *alibi* in existence, the uniqueness of the place I occupy in existence is, in the deepest sense of the word, an answerability.
>
> (Holquist 1990: 30)

Language is dialogic since any discourse involves a dialogic orientation of the utterances of one person to the utterances of others. Any utterance involves at least two voices: the voice producing it and the voice to which it is addressed (Wertsch 1991). An utterance is part of a dialogue because 'any utterance is a link in the chain of speech communication' (Bakhtin 1953: 84). In this sense an utterance responds to previous utterances and anticipates the responses of others. 'Utterances are not indifferent to one another, and are not self-sufficient; they are aware of and mutually reflect one another' (Bakhtin 1953: 91). In addition, any true understanding, or meaning making, is dialogic in nature because we lay down a set of our own answering words for each word of the utterance we are in process of understanding (Voloshinov 1929: 102).

From a dialogic point of view, it is inevitable that a single word has more than one meaning. Indeed, multiplicity of meaning is considered to be a constitutive feature of any word, and the meaning is determined in large part by the prevailing context. Any utterance is considered to be a unique event because of the set of conditions (historical, social, physiological and so on) that are present at the moment the utterance is made and that cannot be reproduced. This does *not* mean that there is no constancy in language. Voloshinov (1929) distinguishes between the *theme* of an utterance and its *meaning*. The *theme* is individual and cannot be reproduced, while its *meaning* accounts for all those aspects of the utterance that are reproducible. One example of the unique nature of the theme of an utterance is given by the *expressive intonation* used by a speaker in populating an utterance with a specific *evaluative accent*. For example, in science classrooms, the students are normally able to recognize whether a question from the teacher is a true question, or simply an evaluation of their understanding; they can do this by paying attention to the intonation of the teacher's voice. In one such situation (Mortimer and Machado 2000), the teacher was discussing with her students the fact that ice floats on water, while a piece of stone sinks. One student suggested that 'the ice is lighter than the stone', and the teacher responded with 'Is the ice *lighter* than the stone?', stressing the word lighter. The student immediately recognized the purpose of the teacher's question and reframed the question 'I mean . . . is the ice less dense than the stone?' Although the teacher's utterance – 'Is the ice *lighter* than the stone' – potentially has a general meaning, which is reproducible in different contexts, its *theme*, acquired in this particular context, was unique.

Meaning and meaning making are governed by two opposing forces: the centripetal, which works to homogenize and to centralize meanings, and the centrifugal, which works to disperse and to decentralize meanings. 'There is a constant interaction between meanings, all of which have the potential of conditioning others' (Holquist 1981: 426). In this sense, language and verbal communication are always dialogic: there can be no monologue.

This final point, that verbal communication must always be dialogic in nature, raises an important issue relating to our use of terms in Chapter 3. In Chapter 3, we use the term *dialogic* as the opposing tendency to *authoritative* in characterizing one of the dimensions of the communicative approach. However, according to Bakhtin's point of view, *all* discourse must be dialogic in form, including authoritative monologue (a non-interactive/authoritative communicative approach). We certainly agree that when a teacher makes a non-interactive authoritative presentation, then the meaning making process is dialogic in nature as the students try to make sense of what is being said by laying down a set of their 'own answering words' to the words of the teacher. At the same time, and according to our own definition, we are clear that in authoritative discourse the teacher's *purpose* is to focus the students' full attention on just *one* meaning. It is in this sense that we have chosen to use the word 'authoritative' (while acknowledging the underlying dialogic nature of the interaction). Additionally, we have chosen the word 'dialogic' to contrast with an authoritative communicative approach, so that we can draw upon the dialogic meaning of recognizing others' points of view. Thus, according to our definition, we are clear that in dialogic discourse the teacher attempts to take into account a range of students', and others', ideas.

The distinction that we make between authoritative and dialogic discourse is in fact based on the distinction that Bakhtin himself makes between *authoritative* and *internally persuasive discourse*. According to Bakhtin:

> Authoritative discourse permits no play with the context framing it, no play with its borders, no gradual and flexible transitions, no spontaneously creative stylizing variants on it. It enters our verbal consciousness as a compact and indivisible mass; one must either totally affirm it, or totally reject it.
>
> (Bakhtin 1934: 343)

Bakhtin exemplifies *authoritative discourse* in terms of: religious dogma, an acknowledged scientific truth, the words of a parent, of teachers and so on. *Internally persuasive discourse*, by contrast, is 'tightly interwoven with "one's own word", it is half-ours and half-someone else's, it represents more than one voice or conceptual horizon' (Bakhtin 1934).

## Note 4: The concept of learning demand

The concept of 'learning demand' (Leach and Scott 2002) offers a way of appraising the *differences* between the social language of school science and the social language that learners brings to the classroom. The purpose of identifying learning demands is to bring into sharper focus the intellectual challenges facing learners as they address a particular aspect of school science. Teaching can then be designed to focus on those learning demands.

An important point relating to the operationalization of the concept of learning demand is that a learning demand can be identified for a *group* of learners working within a specific area of scientific content. This follows from the fact that learners are immersed in a common social language in day-to-day living and will therefore arrive in school with largely similar points of view. In this respect the concept of learning demand is linked more closely to differences between social languages and the meanings that they convey than to differences in the 'mental apparatus' of individuals. Thus, learning demands are *epistemological* rather than *psychological* in nature (Leach and Scott 2002).

### Identifying learning demands

How might learning demands for a particular conceptual area of science be specified? It is possible (Leach and Scott 2002) to identify three ways in which differences between everyday and school science perspectives might arise. These relate to differences in the *conceptual tools* used, differences in the *epistemological underpinning* of those conceptual tools and differences in the *ontology* on which those conceptual tools are based.

For example, in the context of teaching and learning about air pressure, students typically draw upon the everyday concept of *suction* to explain phenomena, while the scientific point of view is based upon differences in air pressure. There is a difference here in the *conceptual tools* used. In relation to plant nutrition, students commonly draw upon everyday notions of *food* as something that is ingested, in contrast with scientific accounts, which describe the synthesis of complex organic molecules within plants from simple inorganic precursors.

Other differences relate to the *epistemological underpinning* of the conceptual tools used. For example, ways of generating explanations using scientific models and theories that are taken for granted in school science are not part of the everyday social language of many learners (Vosniadou 1994; Driver *et al.* 1996). For example, there is evidence that many lower high school students recognize the logical implications of specific pieces of evidence in relation to different models of simple series electrical circuits, but resolve logical inconsistencies by selecting different models to explain the behaviour of different circuits (Leach 1999). The students do not draw upon the epistemological principle of *consistency* that is an important feature of school science. Their everyday social language does not appear to recognize that scientific models and theories ideally explain as broad a range of phenomena as possible.

Learning demands may also result from differences in the *ontology* of the conceptual tools used (Chi 1992; Chi *et al.* 1994; Vosniadou 1994). Thus, entities that are taken for granted as having a real existence in the realm of school science may not be similarly referred to in the everyday language of students. For example, there is evidence that many lower high school students, in learning about matter cycling in ecosystems, do not think about atmospheric gases as a potential source of matter for the chemical processes of ecological systems (Leach *et al.* 1996). There is a learning issue here (of not considering gases to be substantive in nature), which relates to the students' basic commitments about the nature of matter.

## Note 5: Conceptual profiles

The notion of conceptual profile is an attempt to address the problem of how individual meanings are developed in the science classroom, through the interplay between different modes of thinking and ways of speaking (different social languages). The basic assumption is that in any culture, and in any individual, there exists not one, homogeneous form of thinking, but different types of verbal thinking (Tulviste 1991), and that these different modes of thinking are interwoven with different ways of speaking. This heterogeneity of verbal thinking has been characterized in terms of a *conceptual profile* (Mortimer 1995, 1998), which acknowledges the coexistence, for the individual, of two or more meanings for the same word or concept that can be appropriately accessed according to context.

The different ways of thinking and speaking about the world can be thought of as different forms of knowledge that correspond to different views of reality. Among these multiple views of reality, there is one that presents itself as the 'reality par excellence': the reality of everyday life (Berger and Luckmann 1967). When someone moves their attention from this everyday reality to that of scientific knowledge, for example, a radical shift occurs in their consciousness. Nevertheless, even when this kind of radical shift takes place, the reality of everyday life still marks its presence.

Science itself is not a homogeneous form of knowing and speaking and can provide multiple ways of seeing the world that can exist together, in the same individual, and be drawn upon in different contexts. For example, the concept of *the atom* is not restricted to one unique point of view. Chemists deal with the atom as a rigid and indivisible sphere, like the Daltonian atom, to explain several properties of substances. The structural formulae used by chemists also represent the atoms arranged in molecules in this way. This model is not, however, suitable for explaining several phenomena, including chemical reactivity, where more sophisticated models, including those derived from quantum mechanics, are used.

The *conceptual profile* (Mortimer 1995, 1998) provides one means for representing the different ways of conceptualizing reality. These different ways can range from approaches based on everyday knowledge (which might be informed by the immediate sensorial perception of the actual phenomenon) to very sophisticated ways (which might represent reality in purely symbolic models). Between these extremes there are other scientific ways in which the phenomenon might be scrutinized; these include approaches based on empirical experiments or analyses in terms of mathematical relationships between variables. These different ways of making meaning of a word, concept or particular phenomenon can be thought of as constituting different zones of an individual person's conceptual profile.

The concept of mass helps us to understand this conceptual heterogeneity (Bachelard 1968). In our everyday life we normally use the word *mass* or *massive* in referring to big and heavy things. We talk about a *mass of people*, a *mass of bruises*, a *mass of details to be worked out*, a *mass of Sunderland supporters*. In this way, the meaning of the word mass is related strongly to the sense perception of size and heaviness, and it is therefore difficult to think in this way about the mass of entities such as air, gases or electrons.

When we use a balance to determine the mass of objects, our concept of mass becomes related to an *empirical* experience and a precision balance allows us to determine accurately the masses of objects of a few milligrams. If scientists had restricted themselves to this empirical concept of mass, they would not have been able to determine the mass of the Earth, or that of an electron, since there would have been no balance available to do it. To determine the mass of these very big or very small objects, it is necessary to conceptualize mass in a new zone of its profile. Here mass becomes part of a rational relationship between other concepts, which can be expressed as mathematical relationships between variables representing those concepts. In this sense we can think of mass in terms of a relationship between density and volume, or between force and acceleration.

It is possible to extend the profile for mass still further as the relationship between mass and *motion* becomes significant for objects moving at speeds near to the speed of light. It is worth noting that as we go further along the conceptual profile, the concept becomes more complex, in the sense of depending on a greater number of relationships between different concepts.

In terms of the conceptual profile, learning scientific concepts can be thought of as a process involving two distinct but related mental actions:

- acquiring new zones of a specific conceptual profile;
- becoming aware of this changing profile.

For example, a student may development an awareness that the scientific concept of 'heat', as a process of energy transfer ('I heat the water') that depends on the *temperature difference*, is complementary to their everyday concept of heat, which depends on *temperature*. If the notions are complementary, there are contexts in which one of the concepts is more appropriately used than the other. For example, to ask in a shop for a 'warm woollen coat' is far more appropriate than asking for 'a coat made from a good thermal insulator that prevents the body from exchanging heat with the environment'. Furthermore, if we know that the 'warmth' of the wool does not make it suitable for heating water (wool is not a source of heat), we are demonstrating our conscious awareness of this profile, drawing on everyday and scientific ideas in a complementary way.

Each zone in a conceptual profile offers a way of seeing the world that is different from the ways provided by other zones. Each conceptual zone offers different mediational means, different theories and languages that reveal the world in their own way. The real world phenomenon, itself, cannot be understood from just one perspective: only a complementary range of views can give a full picture.

The notion of conceptual profile helps us to operationalize the distinction between *sense* and *meaning* as stated by Vygotsky. According to him:

> A word's sense is the aggregate of all psychological facts that arise in our consciousness as a result of the word. Sense is a dynamic, fluid, and complex formation that has several *zones* that vary in their stability . . . In different contexts, a word's sense changes. In contrast, meaning is a comparatively fixed and stable point, one that remains constant with all the changes of the word's sense that are associated with its use in various contexts . . . Isolated in the lexicon, the word has only one meaning. However, this meaning is nothing more than a potential that can only be realised in living speech, and in living speech meaning is only a cornerstone in the edifice of sense.
>
> (Vygotsky 1934: 275–6)

For Vygotsky sense is something personal, and each word has a different sense for each person. A personal sense for one word is composed of different meanings, which are more stable and fixed. The conceptual profile consists of different zones that correspond to different meanings the word acquires in different contexts. As each person has a different conceptual profile for each concept or word, the profile can be seen as the personal sense of this word for a person. Each zone of the conceptual profile corresponds to a different meaning the word acquires in a specific context.

## Note 6: Zone of proximal development (ZPD)

Vygotsky (1978: 86) defines the zone of proximal development as:

> the distance between the actual developmental level as determined by independent problem solving and the level of potential development as determined through problem solving under adult guidance or in collaboration with more capable peers.

According to Wertsch (1985: 67), Vygotsky introduced this notion 'in an effort to deal with two practical problems: the assessment of children's intellectual abilities and the evaluation of instructional practice'.

The concept of ZPD has become perhaps the most heavily used of Vygotsky's concepts in teaching and learning studies, while, oddly enough, there have been relatively few ZPD-based studies carried out in the specific contexts of assessment and evaluation. An impressive exception to this trend is provided by the work of Brown and Ferrara (1985), who consider the implications of the ZPD for the dynamic assessment of learning potential. Why, then, apart from in the areas of assessment and evaluation, has the ZPD become so popular in studies informed by sociocultural theory?

We see two main reasons. The first is that the ZPD encapsulates the basic principle of Vygotsky's theory concerning the social origins of higher mental functioning. There is an inherent connection between the intermental and intramental planes, as the means that mediate higher mental functions, such as language, are essentially social in nature. Michael Cole (1985: 146) provides a nice expression of this idea, in referring to the ZPD as the place 'where culture and cognition create each other'. The second reason is that the ZPD focuses on the ways in which children might be supported through problem-solving tasks by adults, or through peer collaboration, a theme that has immediate and obvious resonances with teaching and learning. Vygotsky, himself, was the first to signal the potential of the ZPD for thinking about learning, stating that it 'enables us to propound a new formula, namely that the only "good learning" is that which is in advance of development' (Vygotsky 1978: 89).

Bruner (1985), following his work in Wood *et al.* (1976), outlines what has become a very popular way of interpreting the ZPD in teaching and learning situations, through the idea of *scaffolding*:

> If the child is enabled to advance by being under the tutelage of an adult or a more competent peer, then the tutor or the aiding peer serves the learner as a vicarious form of consciousness until such a time as the learner is able to master his own action through his own consciousness and control. When the child achieves that conscious control over a new function or conceptual system, it is then that he is able to use it as a tool. Up to that point, the tutor in effect performs the critical function of 'scaffolding' the learning task to make it possible for the child, in Vygotsky's words, to internalise external knowledge and convert it into a tool for conscious control.
>
> (Bruner 1985: 24–5)

Although the ZPD has become one of the most popular concepts in Vygotsky's work, it is not an idea that is easily put into practice in the context of formal schooling. Many works in the literature refer to the importance of creating ZPDs in classrooms, but the concept was developed to deal with interactions between one adult and one child and classrooms are places where one adult interacts with many children. Although single interactions are played out between the teacher and individual students, all the other students are, in a sense, expected to participate in the same interaction. So how does the concept of ZPD work for a group of students? Since not all the students have the same level of potential development, how might they be guided, or scaffolded, as a whole group? How is the level of assistance required by each student ascertained?

A final point is that, although the concept of the ZPD has been drawn upon and elaborated in many different ways, there are two aspects that appear to be shared by all:

1   A privileging of the intellectual dimension of learning, with a correspondent neglect of affective and emotional influences.
2   A view that encounters between the individual and the other, in situations of learning, are smooth and free of anxiety or tension.

Góes (2001: 84) makes the point that 'This typical characterisation of the inter-subjective functioning has little room for tensions and multiple elaborations that characterise the "joint" activity, and the other is mainly conceived as a participant that helps, shares, guides, builds scaffoldings, etc.' Although the idea of guidance is fundamental to Vygotsky's concept of the ZPD, we agree with Góes (2001: 87) when she argues that: 'If the dynamic of social relations can be tense and conflictive or smooth and co-operative, we cannot think of a prevalent inter-subjective functioning, which implies only part of these characteristics.' In other words, conflict and tension can also characterize 'joint' activity in the ZPD.

## Note 7: Bakhtin's notions of social language and speech genre

Bakhtin assumes that language is never unitary. 'It is unitary only as an abstract grammatical system of normative forms, taken in isolation from concrete, ideological conceptualisations that fill it, and in isolation from the uninterrupted process of historical becoming that is a characteristic of all living languages' (Bakhtin 1934: 288). Language, in the mind of the individual speaker, is not a system of abstract grammatical categories, but a world-view that ensures a maximum of mutual understanding in all spheres of life. In talking or writing, we take our words or phrases not from dictionaries or grammar handbooks, but from the utterances of others. In this sense, Bakhtin identifies the *utterance* as being the fundamental unit of verbal communication.

Bakhtin proposes two forms of stratification of language that ensure its heterogeneity, and these are stratification in terms of *speech genre* and of *social language*. A social language is 'a discourse peculiar to a specific stratum of society (professional, age group, etc.) within a given social system at a given time' (Holquist 1981: 430). All social languages are 'specific points of view on the world, forms for conceptualising the worlds in words, specific world views, each characterised by its own objects, meanings and values . . . As such they encounter one another and co-exist in the consciousness of real people' (Bakhtin 1934: 291–2). In Bakhtin's view, a speaker always produces an utterance using a specific social language that shapes what they can say.

On the other hand, 'a speech genre is not a form of language, but a typical form of utterance; as such the genre also includes a certain typical kind of expression that inheres in it . . . Genres correspond to typical situations of speech communication, typical themes, and, consequently, also to particular contacts between the *meanings* of words and the actual concrete reality under certain typical circumstances' (Bakhtin 1953: 87). Bakhtin gives a list of examples that includes short rejoinders of daily dialogue, everyday narration, the brief standard military command, the elaborate and detailed order, business documents, scientific articles and all literary genres.

Thus, while a social language is related to a specific point of view determined by a social or professional position, the speech genre is related to the social and institutional place where the discourse is produced. For example, in this book we write about the social languages of science and of school science and about the speech genre of classroom talk.

## Note 8: Communicative approach

The term *communicative approach* is used in second language education, and it is important to make clear that we use the term, in this book, with a completely different meaning. This aspect of our analytical framework focuses on questions such as whether or not the teacher interacts with students (taking turns in the discourse), and whether the teacher takes account of students' ideas, as teaching and learning proceed.

In second language education, the communicative approach refers to a strategy for teaching a second language, 'which emphasizes the teaching of communicative functions of language (e.g. requesting, apologizing, disagreeing) rather than of linguistic structures' (Harmer 1991: 3). 'Because of the focus on communicative activities and the concentration on language as a means of communication, such an approach has been called the communicative approach. This is because its aims are overtly communicative and great emphasis is placed on training students to use language for communication' (Harmer 1991: 41).

# Note 9: Empirical and theoretical referents

In order to identify whether a description, explanation or generalization is *empirical* or *theoretical*, it is possible to borrow some concepts from linguistics. The basic notion that we use to make this distinction is that of reference and referentiality. Language has many functions, and to refer to objects, events and abstract ideas is only one of them. Jakobson (1990), for example, identifies six different functions: referential, emotive, poetic, conative, phatic and metalingual. Vygotsky also wrote about the functional diversity of speech. The functions Vygotsky identified can be categorized in pairs, as suggested by Wertsch (1985):

- signalling versus significative;
- social versus individual;
- communicative versus intellectual;
- indicative versus symbolic.

The first term of each pair corresponds to the function that appears early in the development of a child. For example, in referring to the social versus individual pair, Vygotsky noted that 'a sign is always originally a means used for social purposes, a means of influencing others, and only latter becomes a means of influencing oneself' (Wertsch 1985: 92). Referentiality deals with the relationship between a sign (for example, a word) and the object, event or abstract idea that the sign refers to.

Based on the idea that in science lessons people talk about real objects and events, and also about theoretical ideas, we can establish the distinction between empirical and theoretical referentiality and use it to classify descriptions, explanations and generalizations.

If a description is developed in terms of referents visually present in the system, it falls within the first type – empirical. If a description, on the other hand, is based on referents that are not visually present in the system, but are entities created through discourse as part of a symbolic system, this description is of the second type – theoretical. It is important to note that, in the classroom, entities such as electrons, atoms and molecules, independently of their realistic status, are created mainly through discourse. In this sense we refer to such entities as being created by intralinguistic relationships, although science teachers and students tend to consider these entities as real objects. Furthermore, we consider that discussion about whether or not these entities really exist is potentially misleading, as the power of such entities resides not in their real existence, but in the explanatory possibilities that they confer to the theoretical discourse built around them.

In the history of science there are good examples of how an entity can survive as a powerful idea, independently of the experimental evidence of its actual existence. The history of the atomic hypothesis provides one such example. In 1803, John Dalton (1766–1844) presented what is generally accepted as the first atomic theory applied to chemistry. Atomism, at that time, was already well known through the work of Pierre Gassendi (1592–1655) and had been used by several prominent scientists, such as Galileo (1564–1642), Boyle (1627–91) and Newton (1642–1727). Dalton opened a fruitful research programme in chemistry by assuming that weight is a fundamental property of atoms that can be used to distinguish between the atoms of different chemical elements. Nevertheless, throughout the entire nineteenth century, atomism was immersed in controversy and confusion. Not all scientists accepted the Daltonian programme, as Berzelius (1779–1848) did in his search for atomic weights. For example, in 1836, Dumas (1800–84) wrote, 'If I had such power I would erase the word atom from science, persuaded as I am that it goes beyond experience and that in chemistry we should never go beyond experience' (Rheinboldt 1988: 80). Although the existence of atoms was widely accepted only at the beginning of the twentieth century, most of the chemistry of the nineteenth century, including structural and organic chemistry, was already built around the ideas of atoms and molecules well before this consensus had been reached.

## Note 10: Transcribing and translating oral discourse

The focus of this book is on meaning and meaning making in the science classroom. Meaning is viewed as *polysemous* (a single word or utterance can carry more than one meaning) and *polyphonic* (a single word or utterance can be used to express more than one point of view depending on its context).

The approach we have taken to gathering and analysing research data is, itself, viewed as involving a process of meaning making, and this has a methodological consequence. That is, the data are constructed in the interaction and not viewed as something 'out there' that is objectively accessed and analysed. According to Lemke (1998: 1176), 'Because linguistic and cultural meaning, which is what we are ultimately trying to analyse, is always highly context-dependent, researcher-controlled selection, presentation, and re-contextualisation of verbal data are critical determinants of its information content.'

In writing this book we have faced two fundamental difficulties that characterize this area of classroom discourse analysis:

- how to transcribe oral discourse into a written text;
- how to translate the oral discourse produced in one language (Portuguese) into another (English).

According to Jane Edwards (1993: 3), 'the transcript plays a central role in research on spoken discourse, distilling and freezing in time the complex events and aspects of interaction in categories of interest to the researcher.'

We have already mentioned (in Chapter 2) that scientific discourse and school science discourse are multimodal in nature. Classroom talk, as a spoken discourse, is accompanied by features that play an important role in conveying the meanings of the words. Gesture, body movement, pause, changes in gaze direction and facial expression, changes in voice tone and pitch are all examples of such features that accompany spoken discourse (Green and Wallat 1979), and all are difficult to capture in a written transcript. Although we recognize the importance of these features in interpreting discourse, we decided to keep the transcripts in this book as simple as possible in order to prevent the reader from becoming lost in a mass of data. When transcribing data from video records, we normally use additional columns to make a detailed record of these accompanying modes of communication. For the data presented in this book we simply comment on these features within [square brackets] in the body of the transcript when such comments are essential to understanding the meaning of the words. We also *italicize* words in the transcripts where the speaker modulates their voice to make a particular point of emphasis.

In transcribing the classroom discourse, we also decided to arrange the talk in speaker turns, one above the other. According to Ochs (1979), this form of arrangement is not adequate for transcribing conversation between small children because a child's utterance in a dialogue is not necessarily related to the previous utterance. Ochs considers, nevertheless, that it is adequate for adult–adult conversations because there is a mutual interdependence between speakers and an utterance in a dialogue is normally related to the previous one. We consider that the talk between teacher and students, or among students, that we present is this book shares this adult feature.

One further point to be made here relates to the cultural difference between classroom talk in England and Brazil. We have found that the discourse enacted in Brazilian classrooms has far more simultaneous utterances than that in England. Although we could have marked this difference by signalling the simultaneous talk in the transcripts, we decided to avoid reproducing transcripts with a lot of simultaneous talk. Our main aim in using the transcripts was to illustrate the framework, working at an appropriate level of detail of analysis.

A second difficulty we faced in preparing this book was to translate oral Portuguese into written English, a challenging process! There is a huge number of oral expressions in Portuguese that have no English equivalent. In addition, it is impossible to translate into English the abbreviations that are used in oral language, and this issue is further complicated by the existence of regional differences within Brazil. For example, one such abbreviation is peculiar to

Minas Gerais state, where the data were collected. In the sentence 'Você está bom?' the pronoun *você* ('you' in English), is abbreviated to *cê* and the verb 'to be', *está*, to *tá*, generating the abbreviated expression 'Cê tá bom?' A literal translation of this expression to English, word by word, would generate 'Are you OK?' but the meaning of the expression is far closer to 'How are you?' Although it is possible to preserve the meaning in the translation, there is no way to preserve the literal abbreviation.

Being aware of these difficulties, we tried, in translating the classroom talk, to find expressions in English that would be closest to the meanings of the original Portuguese. We also included some English abbreviated expressions in order to project the informal nature of much of the classroom talk. Nevertheless, the reader should be aware that, due to the difficulties outlined above, the transcribed talk taken from the Brazilian classrooms appears rather more formal than it actually is.

Just to provide the reader with one example of what was involved in making these translations, the dialogue we have just referred to in the main text was translated from the following Portuguese:

Cristina: Isto aqui é um líquido, e um líquido não assume forma definida, não é? Ou seja, tem agitação das partículas.
Teacher: Tem agitação das partículas.
Cristina: E existe espaço. Então, como existe agitação e espaço entre as partículas a tendência é que uma não fique no mesmo lugar. Aí, então, elas vão se misturando.

# References

Arruda, S. M., Villani, A. and Laburu, C. E. (2001) Perfil conceitual e/ou perfil subjetivo? In M. A. Moreira, I. M. Greca and S. C. Costa (orgs) *Atas do Terceiro Encontro Nacional de Pesquisa em Educação em Ciências*. Atibaia: ABRAPEC (electronic version on CD-ROM).

Bachelard, G. (1968) *The Philosophy of No*. (trans. G. C. Waterston). New York: The Orion Press.

Bakhtin, M. M. (1934) Discourse in the novel. In *The Dialogic Imagination* (ed. M. Holquist, trans. C. Emerson and M. Holquist). Austin: University of Texas Press (1981).

Bakhtin, M. M. (1953) *Speech Genres and Other Late Essays* (eds. C. Emerson and M. Holquist, trans. V. W. McGee). Austin: University of Texas Press (1986).

Barnes, D. and Todd, F. (1995) *Communication and Learning Revisited: Making Meaning through Talk*. Portsmouth: Boynton/Cook.

Bell, B., with Barron, J. and Stephenson, E. (1985) *The Construction of Meaning and Conceptual Change in Classroom Settings. Case Studies on Plant Nutrition*. Leeds: CSSME, School of Education, University of Leeds.

Berger, P. L. and Luckmann, T. (1967) *The Social Construction of Reality: A Treatise in the Sociology of Knowledge*. London: Allen Lane.

Brown, A. L. and Ferrara, R. A. (1985) Diagnosing zones of proximal development. In J. V. Wertsch (ed.) *Culture, Communication and Cognition: Vygotskian Perspectives*. Cambridge: Cambridge University Press.

Bruner, J. (1985) Vygotsky: an historical and conceptual perspective. In J. Wertsch (ed.) *Culture, Communication and Cognition: Vygotskian Perspectives*. Cambridge: Cambridge University Press.

Chi, M. T. H. (1992) Conceptual change within and across ontological categories: examples from learning and discovery in science. In R. Giere (ed.) *Cognitive Models of Science: Minnesota Studies in the Philosophy of Science*. Minneapolis: University of Minnesota Press.

Chi, M. T. H., Slotta, J. and deLeeuw, N. (1994) From things to processes: a theory of conceptual change for learning science concepts, *Learning and Instruction*, 4: 27–43.

Cole, M. (1985) The zone of proximal development: where culture and cognition create each other. In J. Wertsch (ed.) *Culture, Communication and Cognition: Vygotskian Perspectives*. Cambridge: Cambridge University Press.

Damasio, A. (1994) *Descartes' Error: Emotion, Reason, and the Human Brain*. New York: Avon Books.

Driver, R., Guesne, E. and Tiberghien, A. (1985) *Children's Ideas in Science*. Milton Keynes: Open University Press.

Driver, R., Leach, J., Millar, R. and Scott, P. (1996) *Young People's Images of Science*. Buckingham: Open University Press.

Driver, R., Newton, P. and Osborne, J. (2000) Establishing the norms of scientific argumentation in classrooms, *Science Education*, 84(3): 287–312.

Duschl, R. A. and Osborne, J. (2002) Supporting and promoting argumentation discourse in science education, *Studies in Science Education*, 38: 39–72.

Edwards, D. and Mercer, N. (1987) *Common Knowledge: The Development of Understanding in the Classroom*. London: Routledge.

Edwards, J. A. (1993) Principles and contrasting systems of discourse transcription. In J. A. Edwards and M. D. Lampert (eds) *Talking Data: Transcription and Coding in Discourse Research*. Hillsdale, NJ: Lawrence Erlbaum Associates.

Góes, M. C. R. (2001) A construção de conhecimentos e o conceito de zona de desenvolvimento proximal. In E. F. Mortimer and A. L. B. Smolka (eds) *Linguagem, cultura e cognição: reflexões para o ensino e a sala de aula*. Belo Horizonte: Autêntica.

Green, J. and Wallat, C. (1979) What is an instructional context? An exploratory analysis of conversational shifts across time. In O. Garnica and M. King (eds) *Language, Children and Society*. New York: Pergamon.

Harmer, J. (1991) *The Practice of English Language Teaching*. London: Longman.

Holquist, M. (1981) *Bakhtin, M. M. The Dialogic Imagination*. Austin: University of Texas Press.

Holquist, M. (1990) *Dialogism: Bakhtin and His World*. New York: Routledge.

Jakobson, R. (1990) *On Language*. Cambridge, MA: Harvard University Press.

Johnston, K. and Driver, R. (1991) *A Constructivist Approach to the Teaching of the Particulate Theory of Matter: A Report on a Scheme in Action*. Leeds: Children's Learning in Science Project, CSSME, School of Education, University of Leeds.

Josephs, I. E. (1998) Do you know Ragnar Rommetveit? On dialogue and silence, poetry and pedantry, and cleverness and wisdom in psychology (an interview with Ragnar Rommetveit), *Culture and Psychology*, 4(2): 189–212.

Kelly, G. J., Brown, C. and Crawford, T. (2000) Experiments, contingencies and curriculum: providing opportunities for learning through improvisation in science teaching, *Science Education*, 84(5): 624–57.

Kress, G., Jewitt, C., Ogborn, J. and Tsatsarelis, C. (2001) *Multimodal Teaching and Learning: The Rhetorics of the Science Classroom*. London: Continuum.

Leach, J. (1999) Students' skills in the co-ordination of theory and evidence in science, *International Journal of Science Education*, 21(8): 789–806.

Leach, J., Ametller, J., Hind, A., Lewis, J. and Scott, P. (2003) Evidence-informed approaches to teaching science at junior high school level: outcomes in terms of student learning. Paper presented at the Annual Meeting of the National Association for Research in Science Teaching, Philadelphia, March.

Leach, J., Driver, R., Scott, P. and Wood-Robinson C. (1996) Children's ideas about ecology 2: ideas about the cycling of matter found in children aged 5–16, *International Journal of Science Education*, 18(1): 19–34.

Leach, J. and Scott, P. (2002) Designing and evaluating science teaching sequences: an approach drawing upon the concept of learning demand and a social constructivist perspective on learning, *Studies in Science Education*, 38: 115–42.

Lemke, J. L. (1990) *Talking Science: Language, Learning and Values*. Norwood, NJ: Ablex Publishing Corporation.

Lemke, J. L. (1998) Analysing verbal data: principles, methods, and problems. In B. J. Fraser and K. G. Tobin (eds) *International Handbook of Science Education*. Dordrecht: Kluwer Academic Publishers.

Mehan, H. (1979) *Learning Lessons: Social Organization in the Classroom*. Cambridge, MA: Harvard University Press.

Millar, R. and Osborne, J. (1999) *Beyond 2000*. London: King's College.

Mortimer, E. F. (1995) Conceptual change or conceptual profile change?, *Science and Education*, 4: 267–85.

Mortimer, E. F. (1998) Multivoicedness and univocality in the classroom discourse: an example from theory of matter, *International Journal of Science Education*, 20(1), 67–82.

Mortimer, E. F. (2000) *Linguagem e formação de conceitos no ensino de ciências*. Belo Horizonte: Editora UFMG.

Mortimer, E. F. and Machado, A. H. (1996) A linguagem numa aula de ciências, *Presença Pedagógica*, 2(11): 49–57.

Mortimer, E. F. and Machado, A. H. (2000) Anomalies and conflicts in classroom discourse, *Science Education*, 84: 429–44.

Mortimer, E. F. and Scott, P. H. (2000) Analysing discourse in the science classroom. In J. Leach, R. Millar and J. Osborne (eds) *Improving Science Education: The Contribution of Research*. Buckingham: Open University Press.

Ochs, E. (1979) Transcription as theory. In E. Ochs and B. B. Schieffelin (eds) *Developmental Pragmatics*. New York: Academic.

Ogborn, J., Kress, G., Martins, I. and McGillicuddy, K. (1996) *Explaining Science in the Classroom*. Buckingham: Open University Press.

Posner, G. J., Strike, K. A., Hewson, P. W. and Gerzog, W. A. (1982) Accommodation of a scientific conception: toward a theory of conceptual change, *Science Education*, 66(2): 211–27.

Rheinboldt, H. (1988) *História da Balança. A vida de J. J. Berzelius*. São Paulo, Nova Stella/EDUSP.

Roychoudhury, A. and Roth, W. M. (1996) Interactions in an open-inquiry physics laboratory, *International Journal of Science Education*, 18(4): 423–45.

Scott, P. (1997) Developing science concepts in secondary classrooms: an analysis of pedagogical interactions from a Vygotskian perspective. Unpublished PhD dissertation, University of Leeds.

Scott, P. (1998) Teacher talk and meaning making in science classrooms: a Vygot-skian analysis and review, *Studies in Science Education*, 32: 45–80.

Scott, P., Hind, A., Leach, J. and Lewis, J. (2001) Designing and implementing science teaching drawing upon research evidence about science teaching and learning. Paper presented at the European Science Education Research Association (ESERA), Third International Conference, Thessaloniki, Greece, 21–5 August.

Sutton, C. R. (1996) The scientific model as a form of speech. In G. Welford, J. Osborne and P. Scott (eds) *Research in Science Education in Europe*. London: Falmer Press.

Toulmin, S. (1978) The Mozart of psychology, *New York Review of Books*, September.

Tulviste, P. (1991) *The Cultural–Historical Development of Verbal Thinking* (trans. M. J. Hall). Commak, NY: Nova Science.

Voloshinov, V. N. (1929) *Marxism and the Philosophy of Language*. Cambridge, MA: Harvard University Press (1973).

Vosniadou, S. (1994) Capturing and modelling the process of conceptual change, *Learning and Instruction*, **4**: 45–69.

Vygotsky, L. S. (1931) The genesis of higher mental functions. In J. V. Wertsch (ed.) *The Concept of Activity in Soviet Psychology*. Armonk, NY: M. E. Sarpe (1981).

Vygotsky, L. S. (1934) Thinking and speech. In *The Collected Works of L. S. Vygotsky* (eds. R. W. Rieber and A. S. Carton, trans. N. Minich). New York: Plenum Press (1987).

Vygotsky, L. S. (1978) *Mind in Society: The Development of Higher Psychological Processes*. Cambridge, MA: Harvard University Press.

Wertsch, J. V. (1985) *Vygotsky and the Social Formation of Mind*. Cambridge, MA: Harvard University Press.

Wertsch, J. V. (1991) *Voices of the Mind: A Sociocultural Approach to Mediated Action*. Cambridge, MA: Harvard University Press.

Wightman, T. (1986) *The Construction of Meaning and Conceptual Change in Class-room Settings: Case Studies in the Particulate Theory of Matter*. Leeds: Children's Learning in Science Project, CSSME, School of Education, University of Leeds.

Wood, D. J., Bruner, J. S. and Ross, G. (1976) The role of tutoring in problem solving, *Journal of Psychology and Psychiatry*, **17**: 89–100.

# Index

# LANGUAGE AND LITERACY IN SCIENCE EDUCATION

## Jerry Wellington and Jonathan Osborne

All teachers look and hope for more scientific forms of expression and reasoning from their pupils, but few have been taught specific techniques for supporting students' use of scientific language. This book is full of them . . . In this very practical book, Jerry Wellington and Jonathan Osborne do much more than summarize research which shows how very much language, in all its forms, matters to science education. They also show teachers what can be done to make learning science through language both more effective and more enjoyable.

Jay L. Lemke, Professor of Education, City University of New York

Science in secondary schools has tended to be viewed mainly as a 'practical subject', and language and literacy in science education have been neglected. But learning the language of science is a major part of science education: every science lesson is a language lesson, and language is a major barrier to most school students in learning science. This accessible book explores the main difficulties in the language of science and examines practical ways to aid students in retaining, understanding, reading, speaking and writing scientific language.

Jerry Wellington and Jonathan Osborne draw together and synthesize current good practice, thinking and research in this field. They use many practical examples, illustrations and tried-and-tested materials to exemplify principles and to provide guidelines in developing language and literacy in the learning of science. They also consider the impact that the growing use of information and communications technology has had, and will have, on writing, reading and information handling in science lessons.

The authors argue that paying more attention to language in science classrooms is one of the most important acts in improving the quality of science education. This is a significant and very readable book for all student and practising secondary school science teachers, for science advisers and school mentors.

### Contents

*Acknowledgements – Introduction: the importance of language in science education – Looking at the language of science – Talk of the classroom: language interactions between teachers and pupils – Learning from reading – Writing for learning in science – Discussion in school science: learning science through talking – Writing text for learning science – Practical ploys for the classroom – Last thoughts . . . – References – Appendix – Index.*

160pp    0 335 20598 4 (Paperback)    0 335 20599 2 (Hardback)

# IMPROVING SCIENCE EDUCATION
## THE CONTRIBUTION OF RESEARCH
### Robin Millar, John Leach and Jonathan Osborne (eds)

This book takes stock of where we are in science education research, and considers where we ought now to be going. It explores how and whether the research effort in science education has contributed to improvements in the practice of teaching science and the science curriculum. It contains contributions from an international group of science educators. Each chapter explores a specific area of research in science education, considering why this research is worth doing, and its potential for development. Together they look candidly at important general issues such as the impact of research on classroom practice and the development of science education as a progressive field of research. The book was produced in celebration of the work of the late Rosalind Driver. All the principal contributors to the book had professional links with her, and the three sections of the book focus on issues that were of central importance in her work: research on teaching and learning in science; the role of science within the school curriculum and the nature of the science education we ought to be providing for young people; and the achievements of, and future agenda for, research in science education.

*Contents*

384pp    335 20645 X (Paperback)    0 335 20646 8 (Hardback)